T0140754

Providing orbit information with predetermined bounded accuracy

Von der Fakultät für Maschinenbau
der Technischen Universität Carolo-Wilhelmina zu Braunschweig

zur Erlangung der Würde

eines Doktor-Ingenieurs (Dr.-Ing.)

genehmigte Dissertation

von: Vitali Braun
aus: Pleschanowo

eingereicht am: 04.04.2016
mündliche Prüfung am: 26.08.2016

Gutachter: Prof. Dr.-Ing. Enrico Stoll, B. Sc.
Prof. Dr.-Ing. Heiner Klinkrad
Prof. Dr.-Ing. Joachim Block
Ph.D. Moriba K. Jah

2016

Bibliografische Information der Deutschen Nationalbibliothek

Die Deutsche Nationalbibliothek verzeichnet diese Publikation in der Deutschen Nationalbibliografie; detaillierte bibliografische Daten sind im Internet über http://dnb.d-nb.de abrufbar.

ISBN 978-3-8325-4405-8

Logos Verlag Berlin GmbH
Comeniushof, Gubener Str. 47,
10243 Berlin
Tel.: +49 (0)30 42 85 10 90
Fax: +49 (0)30 42 85 10 92
INTERNET: http://www.logos-verlag.de

Abstract

The exchange of orbit information is becoming more important in view of the increasing population of objects in space as well as the increase in parties involved in space operations. The US Space Surveillance Network is an example for a system which obtains orbit data from measurements provided by a network of globally distributed sensors.

The aim of this thesis was to highlight how the orbit information maintained by a surveillance system is provided to the users of such a system. Services like collision avoidance require very accurate information, while other services might work with less accurate data. Individual users or entities might have different privileges concerning data they are able to access.

An approach was studied, which allows to derive orbit information of predetermined accuracy from a reference orbit. The method is based on a least-squares fit with a modified geopotential. It was shown that the method works for any Earth orbit and also provides the residuals with respect to the reference, that can be used to construct a covariance matrix in addition to the state vector information.

While orbit information today is often provided via tables of time-tagged ephemerides, an approach was studied to use Chebyshev polynomials that allow for the provision of continuous state vector and covariance matrix information. The major advantage is that a user of this data does not have to do any extrapolation on his own and thus directly retrieves the object's orbit as provided for a given time span.

Representing the orbit in terms of series of polynomial coefficients is referred to as ephemeris compression. It was shown, that the compression rates can be very high. Moreover, a method to also reduce the data amount by interpolating the variance envelope functions was studied.

The method proposed in this thesis to provide state vector and covariance matrix information of predetermined accuracy, gives access to highly accurate information from the catalogue, where this information is required. On the other hand it can also provide less accurate information, where the requirements are less restrictive, thereby allowing for a significantly reduced amount of data to be transferred and stored.

Kurzfassung

Der Austausch von Bahndaten gewinnt zunehmend an Bedeutung: sowohl im Hinblick auf die steigende Population an Objekten in erdgebundenen Bahnen, als auch die zunehmende Zahl an beteiligten Parteien im Betrieb von Raumfahrtmissionen. Das amerikanische SSN (Space Surveillance Network) ist ein Beispiel für ein System, welches Bahndaten aus den Beobachtungen eines global verteilten Sensornetzwerks gewinnt.

In dieser Arbeit wird untersucht, wie Bahndaten eines solchen Systems dessen Nutzern zur Verfügung gestellt werden. Darauf basierende Dienstleistungen, wie etwa die Kollisionsvermeidung, stellen hohe Ansprüche an die Genauigkeit der katalogisierten Bahnen. Andererseits gibt es Dienste, die mit deutlich geringeren Genauigkeiten zuverlässig funktionieren. Individuelle Nutzer, Gruppen oder auch staatliche Einrichtungen können unterschiedliche Privilegien im Hinblick darauf besitzen, zu welchen Daten und welcher Genauigkeit sie Zugang erhalten.

Eine Methode zur Ableitung von Bahndaten mit vorab festgelegter Genauigkeit im Vergleich zu einer Referenztrajektorie wird untersucht. Dieser Ansatz basiert auf der Methode der kleinsten Quadrate und sucht eine Bahnlösung mit modifizierten Störtermen für das Geopotential. Die Residuen einer solchen Lösung gegenüber der Referenzbahn konnten auch genutzt werden, eine kombinierte Kovarianzmatrix neben der Bahnlösung zu ermitteln.

Während Bahndaten heute meist in Form von tabulierten Ephemeriden ausgetauscht werden, wird in dieser Arbeit ein Ansatz mit Tschebyschow-Polynomen verfolgt. Diese erlauben es, Bahndaten und zugehörige Unsicherheiten in kontinuierlicher Form bereitzustellen. Ein großer sich daraus ergebender Vorteil ist, dass ein Nutzer der Daten keine bahnmechanischen Extrapolationen benötigt und für ein gegebenes Zeitintervall direkt die Bahnlösung erhält.

Darüber hinaus wird ein Ansatz diskutiert, mit dem die Varianzen der Komponenten des Zustandsvektors eines Objekts in der Zeit vorwärts gerechnet werden, anschließend eine Einhüllende für diese berechnet und Letztere dann interpoliert wird.

Die in dieser Arbeit vorgestellte Methode, Bahndaten und zugehörige Unsicherheiten in verschiedenen Genauigkeitsklassen zur Verfügung zu stellen, verspricht Nutzern hochgenaue Daten, wenn diese benötigt werden. Auf der anderen Seite ermöglicht sie eine deutliche Reduktion des Datenaufkommens, insbesondere für Dienstleistungen und Nutzer, die vergleichsweise niedrige Anforderungen an die Bahngenauigkeit stellen.

Preface

In 2008, the European Space Agency (ESA) initiated the Space Situational Awareness (SSA) programme, which was declared as an optional programme at the Ministerial Council in November 2008. The first phase was launched in 2009 focusing on three areas: Space Weather (SWE), Near-Earth Objects (NEO) and Space Surveillance and Tracking (SST). The objective was to create a system made up of European assets which provide timely and reliable data, information and services regarding the space environment.

During the Preparatory Programme between 2009 and 2012, ESA and the Institute of Space Systems at the TU Braunschweig launched a cooperation in the frame of ESA's Networking/Partnering Initiative (NPI) to develop core SST technologies related to orbit and covariance data formats and catalogue backend processing. With a substantial background in space debris research, the TU Braunschweig had all the prerequisites to significantly contribute to ESA's goals in the SST programme.

I was selected as the doctoral student in this NPI in 2011. Only knowing the very basics of orbital mechanics back then, I was fascinated to learn about predicting trajectories of objects in space and contribute to a European space surveillance system. When I started outlining my research, I realised very soon that my work would be covering many different areas, including numerical analysis, software engineering, orbit determination and estimation theory. This is why I felt very honoured to be supervised by Prof. Heiner Klinkrad, the Head of ESA's Space Debris Office at that time and an outstanding expert in this research area. I am deeply grateful for all the discussions we had, the critical reflection of every aspect of this work and the continuous support for more than five years.

I am grateful to Prof. Peter Vörsmann and Dr. Carsten Wiedemann, who gave me the chance to become part of the Space Debris working group at the TU Braunschweig in 2010. After the retirement of Prof. Vörsmann, I felt delighted being accepted as a doctoral student by his successor as the head of the institute, Prof. Enrico Stoll, when my work in late 2014 was already in a very mature state. I have to express my sincere thanks to Prof. Stoll for his support and the feedback on verious technical but also organisational aspects.

Every scientific contribution needs to be critically reviewed by experts in the field. It was important for me to get feedback also from scientists external to the NPI between TU Braunschweig and ESA. I want to acknowledge Moriba Jah, Associate Professor and director of the Space Object Behavioral Sciences at the University of Arizona, from whom I learned even more about my subject. I am very grateful for the time he dedicated to review this work.

I have to express my sincere thanks to Prof. Joachim Block and Prof. Rinie Akkermans for being in my doctoral committee.

There are many more people I met, discussed with, learned and even received contributions from during the past few years. I have to say thank you to my students Sebastian Weidemeyer, Michael Baade, Guido Lange, Esfandiar Farahvashi and André Horstmann, who provided valuable contributions to the NEPTUNE software. I had many opportunities to reflect on various details of my work with my colleagues at the university, and I have to thank Sven Flegel, Johannes Gelhaus, Marek Möckel, Christopher Kebschull for their time. At ESA's Space Debris Office, I received a lot of feedback and I have to say thank you to Jan Siminski, Holger Krag, Tim Flohrer for their support. Many thanks go to all the people from TU Braunschweig, ESA, to my friends, whom I did not mention here. I know, I owe you a lot.

There are no words to describe that special feeling you have when a child is born. Levin and Milan came into my life during the time I was working on this thesis and they taught me the most important lesson. Seeing how life evolves from the very first moment changes everything - never before did I feel this great appreciation for life. Eugenia, thank you for making this all possible. Thank you for your love, your patience and your never-ending support!

The support of my parents and my brother has been invaluable to me. Thank you for always being there for me!

Contents

1

Introduction

With the onset of spaceflight operations, the provision of data and information on objects residing in the near-Earth space environment has been of ever-growing importance. The US Space Surveillance Network (SSN) is the most comprehensive combination of globally distributed sensors to observe, track and catalogue the on-orbit object population. Tasked by the Joint Space Operations Center (JSpOC) under the United States Strategic Command (USSTRATCOM), this network is currently keeping track of more than 17 800[1] catalogued objects.

For many decades, information on individual objects has been provided in a so called Two Line Elements (TLE) format to users worldwide. Specifically designed for the purpose of tracking space objects, TLE are based on a simplified general perturbations (GP) theory resulting in low computational requirements. However, due to the need of high accuracy information in various practical applications, like geodesy or oceanography, an individual satellite's owner and/or operator (O/O) would collect his own measurements to augment the available data.

Satellites operated in the densely populated orbits between 600 km to 1000 km are routinely performing collision avoidance manoeuvres. For example, in the year 2013, there were seven conjunctions with a miss distance of less than 300 m for ESA's Cryosat-2 which led to two evasive manoeuvres (Klinkrad, 2014).

Until 2009, conjunction analysis was performed by the O/O through comparing their accurate solutions for the satellite they were responsible for, with the trajectories of objects from the TLE catalogue. The main problem in this process is that TLE data do not include any uncertainty measure for the provided state information. In addition, the GP theory results in typical position errors in the Low Earth Orbit (LEO) region in the order of magnitude of 100 m (Flohrer et al., 2008) at TLE epoch, with the largest error being in the direction of motion.

For the Space Shuttle and the International Space Station (ISS), NASA performed conjunction analysis based on Orbital Conjunction Messages (OCMs) issued by USSTRATCOM. Those messages contained uncertainty information for both, target and risk objects, which were derived from Special Perturbations (SP) techniques, known to provide

[1] https://www.space-track.org, as of December 5, 2016

accurate numerical orbit predictions. However, the OCM were not available to non-US entities.

The accidental collision between the two intact spacecraft Cosmos-2251 and Iridium-33 on February 10, 2009, in retrospect, may be considered as a pivotal incident in the way this process changed. In the aftermath of this event, leading to 1668 and 628 catalogued fragments[2] for Cosmos-2251 and Iridium-33, respectively, an internal review of the conjunction assessment process at USSTRATCOM was performed, which ultimately resulted in a law allowing USSTRATCOM to share Space Situational Awareness (SSA) data with non-US entities. In 2010, USSTRATCOM began providing collision warnings to their partners via so-called Conjunction Summary Messages (CSMs) containing state vector and uncertainty information for both objects at the time of closest approach.

Providing information based on SP techniques significantly improved the collision risk estimation process. However, the CSM, later being subject to a data format standardization process and becoming to what is today known as the Conjunction Data Message (CDM), contained only information for the individual conjunction events. With the Space Surveillance Network (SSN) being a military observation network, users and O/O are still restricted to using TLE data or own observations for general tasks, while being provided with accurate SP data in support of conjunction assessment only. The USSTRATCOM is thus "walking that line between transparency and security" (Bird, 2010).

In 2009, the European Space Agency (ESA) launched its SSA programme with the main objective to "support Europe's independent utilisation of, and access to, space through the provision of timely and accurate information and data [...]"[3]. Being an intergovernmental organization, a European observation network operated by ESA would inevitably have to implement a data sharing policy between its member states and individual users, as again, like for the JSpOC-tasked SSN, sensitive information might be collected, interfering with national security policies and thereby impeding an open data exchange between the member states.

1.1. Motivation and scope

One of the key questions for any SSA system is related to how and to what extent collected data should be provided to its users and customers. With different sensors involved in taking observations, a space surveillance network is in general a heterogenous system concerning the amount and quality of data obtained at the different sites. Also, a subset of the derived information from the observations is classified and thus not to be disclosed to the public.

Taking a look at how JSpOC is providing information on in-orbit objects, we can see that while there is a SP catalogue containing the high accuracy data of all objects, that catalogue is, in general, not accessible. In order to provide orbit information, all orbits

---

[2]as of December 5, 2016.

[3]`http://www.esa.int/Our_Activities/Operations/Space_Situational_Awareness/About_SSA`, as of December 5, 2016

are subject to a fitting process, which applies an analytical model to ultimately provide TLE data to the public (Hejduk et al., 2013; McKissock, 2016).

The problem is that assessing the accuracy associated with TLE and their underlying analytical model is very difficult. The motivation for this thesis is to extend the already operational methodology of having a non-public high-accuracy catalogue, with the possibility of deriving orbit information with an assessment of the associated uncertainties. A method is devised, which allows to have more flexibility in pre-defining the accuracy of the generated orbit product, as opposed to having only one solution when deriving TLE.

With orbit information of a given accuracy being available, one key question is which format to use to distribute that information. In the recent years, orbit data messages have been standardised and provide great flexibility in representing all kind of information associated with an on-orbit object.

Furthermore, having orbit information available, users require methods for interpolation, as data typically is provided as a set of discrete points referred to as ephemerides. An approach to provide continuous data would be beneficial, when users do not require a dedicated software package to recover orbit information at any point in time. This was another point which motivated this dissertation and a method will be presented, which not only allows to provide continuous data, but is also compatible with standardised data messages. In Figure 1.1 a summary is given, showing the information flow from the catalogue to the user in the conventional way compared with a method which represents the motivation for this thesis.

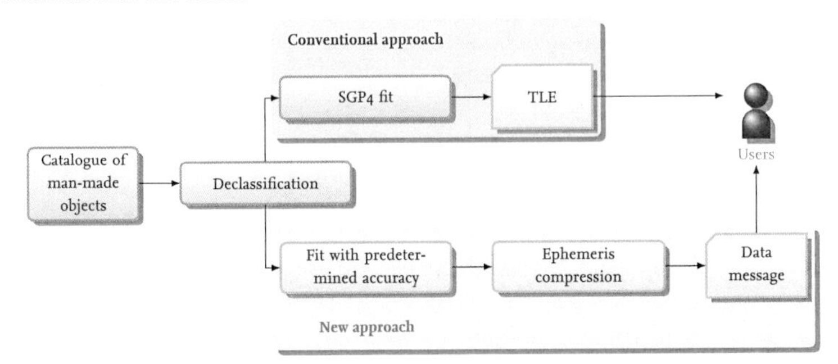

Figure 1.1.: Data flow from the object catalogue to a general user. The *conventional* approach is shown for the provision of TLE data, which are the result of fitting the SP ephemeris with the analytical SGP4 theory (Hejduk et al., 2013; McKissock, 2016). The new approach gives the motivation for this thesis, including an orbit fit with predetermined accuracy and a subsequent interpolation, the latter referred to as *ephemeris compression*.

In the following section, the state-of-the art of collecting and distributing orbital information is described, including systems that have been operational for many decades, as well as the ambitious efforts to establish an SSA system in Europe. This provides some insight into who the users of such a system are and which services there are relying on orbit information of differing quality. Based on this, the key questions this thesis shall adress will be defined in Section 1.3.1, and the methodology is outlined thereafter.

1.2. Background

Comprehensive information on objects orbiting Earth is essential to a wide spectrum of different users and applications. Hence, in order to obtain the required data, dedicated sensors are required, which are typically operated within a network of globally distributed stations.

So called space surveillance systems shall be first described in the following paragraph in order to understand how observational data is obtained. In the next step, it will be possible to understand the orbital data products which are to be provided to certain groups of users. Those users as well as their requirements shall be specified subsequently.

Operational space surveillance systems

The first space surveillance systems, serving to detect, track, catalogue and identify objects orbiting Earth, were devised and implemented in the context of the first few launches. With military users having a key interest in such systems becoming operational, the first ground-based network of sensors, placed at more than 150 different sites, was controlled by the United States (US) via the National Space Surveillance Control Center (NSSCC) from the late 1950's (Hoots et al., 2004). The first version of the Satellite Catalog served to support ballistic missile early warning systems operated by the US Air Force, as well as to alert US Navy fleet units of being observed by reconnaissance satellites (Hoots et al., 2004). Gradually, the sensor network and the control structures evolved to become what is known today as the US SSN operated by JSpOC, the latter being commanded by the USSTRATCOM. The SSN is tasked by JSpOC to collect between 380 000 and 420 000 observations each day (U.S. Strategic Command, 2015) and keep track of more than 17 800 objects.

Today's workhorse of the SSN is the AN/FPS-85 phased array radar system at Eglin AFB (see Figure 1.2), after another main contributor, the Air Force Space Surveillance System (AFSSS), also known as the *Space Fence*, had been shutdown on September 1, 2013. The AN/FPS-85 accounts for 30 % of the total workload of the SSN and can detect, track and identify up to 200 satellites simultaneously (PAFB, 2015). This task is accomplished by its scan coverage of 120° in azimuth and from +3° to +105° in elevation, with the antenna beam pointing south and being inclined 45° with respect to the local horizon (PAFB, 2015), which allows to track up to 95 % of all catalogue objects (Klinkrad, 2006).

While phased array systems operated in surveillance mode would typically be targeting at objects in low Earth orbits (although, for AN/FPS-85, objects greater than about 25 cm can be detected up to a distance of 22 000 nautical miles (PAFB, 2015)), passive optical systems are used to observe deep space objects. Within the SSN, the Ground-based Electro-Optical Deep-Space Surveillance (GEODSS), the Maui Optical Tracking and Identification Facility (MOTIF), as well as the newly installed Space Surveillance Telescope (SST) in 2011, are operational sites used for observing satellites and space debris at higher altitudes. The

(a) Part of the master transmitter antenna at Lake Kickapoo, Texas, contributing to the AFSSS until 2013. (Public domain image, https://commons.wikimedia.org)

(b) The AN/FPS-85 phased array radar at Eglin AFB, Florida. (Public domain image, https://commons.wikimedia.org)

Figure 1.2.: Until September 2013, the workhorses of the SSN had been the AFSSS (a), also known as the Space Fence, and the phased array at Eglin AFB (b). The discontinuation of the Space Fence operation involved modifying operational modes at Eglin AFB in order to maintain routine SSA operations.

(a) The Diego Garcia site, as one of three GEODSS sites, consists of three Cassegrain telescopes and is located in the Indian ocean. (Public domain image, https://commons.wikimedia.org)

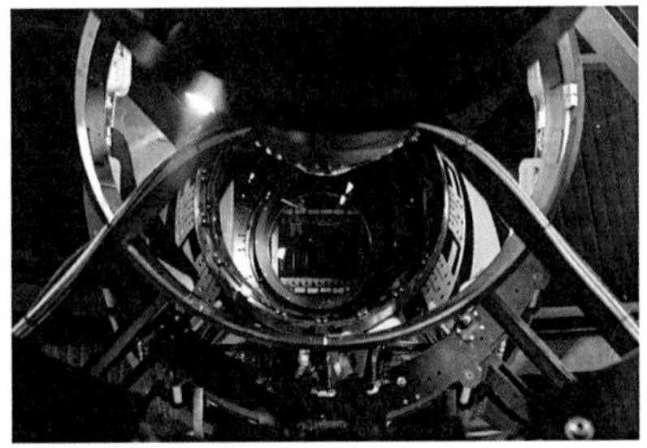

(b) The new Space Surveillance Telescope (SST), developed by DARPA, has undergone testing in the last few years and shall become operational in 2016, supporting the SSN from Australia. (Image source: DARPA)

Figure 1.3.: A few examples of optical systems contributing to the SSN.

GEODSS sites are distributed around the globe to enable observations, for example, of the whole geostationary region. Some of the dedicated optical sensors are shown in Figure 1.3.

A major drawback of ground-based optical systems is that observations can only be taken at night and under clear weather conditions. In order to sidestep these restrictions, the US Air Force is developing the Space Based Space Surveillance (SBSS) system, which is a space-based constellation of optical sensors. In 2010, the first SBSS satellite was launched, carrying a 30 cm telescope (Ball Aerospace, 2015) ought to provide observations of all geosynchronous objects from its sun-synchronous orbit.

Concurrently with the development of the US SSN, the Soviet Union began with its own space surveillance programme from the early 1960's. The (Russian) Space Surveillance System (*Sistema kontrolja kosmicheskogo prostranstva*, SKKP) has been operated by the military from the early days, the motivation being basically the same as for the US system: to support ballistic missile defence operations (Gavrilin, 2008). Besides the various radar sites on the territory of the former Soviet Union, which are part of the Early Warning System (EWS) network and of *Dnestr*, *Dnepr* or *Daryal* type, there are also dedicated space surveillance sensors. Being operated by the 821st Main Space Surveillance Centre, the most important facilities Okno ("window") and Krona ("crown", radioopticheskу kompleks raspoznavaniya kosmicheskikh obektov) were both built in 1999. While the Krona complex in the Caucasus employs both, optical and radar means, the Okno facility in Tajikistan consists of multiple telescopes to monitor objects at altitudes above LEO.

Assessing the capabilities of the surveillance system is not easy and provided figures are quite differing. For example, Allahdadi et al. (2013) states that the number of tracked objects in the Russian catalogue is about 5000. A similar number is provided on the website of GlobalSecurity.org (2015). However, considering the fact that with Krona and Okno two relatively new systems were added to the network, with the latter specifically designed for deep-space observations, and noting a statement of a Russian Colonel (A. Nestechuk) from 2011, stating that four new radar systems will be added to the network till 2020[4], it can be argued that the system clearly evolved in the recent years and will continue to do so, although sensor sites at lower latitudes are still missing. Interestingly, Nestechuk also gives the number of tracked objects as 12 000, which clearly differs from the numbers given by the other authors mentioned above.

In contrast to the US system, the data gathered by the Russian network is for military users only and can not be accessed publicly. However, one can get a quite good impression of the orbit theory behind the cataloguing by referencing, for example, Khutorovsky (2007) or Boikov et al. (2009).

While currently only the USA and Russia have comprehensive space surveillance capabilities, there are many other sensors and networks, like the International Scientific Optical Network (ISON), or national radar facilities in different countries, providing individual observations, tracking support and debris research possibilities, for example, the German Tracking and Imaging Radar (TIRA) or the European Incoherent Scatter Scientific Association (EISCAT).

The situation in Europe

In Europe, agencies and other users so far have been largely dependent on orbit data, for different applications, provided by USSTRATCOM. The situation started to change in the 1990's, when the French Department of Defense (DoD) started to work on the Grand

[4]`http://www.gazeta.ru/social/2011/09/21/3776721.shtml`, accessed on January 7, 2015.

Réseau Adapté a la Veille Spatiale (GRAVES) system, which became operational in 2001 (Klinkrad, 2006). It is able to perform space surveillance tasks with its phased array transmitters at Dijon and a receiver array, based on Yagi antennae, at Apt and was able to catalogue 2200 objects during a one month test in 2001 (Klinkrad, 2006).

In 2009, ESA launched its Space Situational Awareness Preparatory Programme (SSA-PP), with its overall aim being "to support the European independent utilisation of and access to space for research services, through providing timely and quality data, information, services and knowledge regarding the environment, the threats and the sustainable exploitation of the outer space surrounding our planet Earth" (ESA Council, 2008). With the ever increasing number of satellites on orbit and services on Earth relying on the space infrastructure, like weather or navigation applications, the European SSA, with the latter being defined as "a comprehensive knowledge, understanding and maintained awareness of the (i) population of space objects, of the (ii) space environment, and of the (iii) existing threats/risks" (ESA Council, 2008), is aiming towards all kinds of user groups, with the military being only one of them. In 2012, the mandate was extended until 2019, going from the preparatory phase into the so-called Phase II, which puts increased emphasis on the two branches *Space Weather* (SWE) and *Near-Earth Objects* (NEO).

Contemporaneously, the European Union (EU) has started to promote the development of an SSA system through its member states. Recognizing the member states' national assets, like TIRA or GRAVES, being associated with national security requirements, an exemplary work was the Support to Precursor Space Situational Awareness Services (SPA) project under the European Union's Seventh Framework Program (FP7) studying aspects of SSA governance and data policy (Valero et al., 2013). As Valero et al. (2013) point out, the principle behind SSA governance and data policy lies in protecting "the interests of the EU, its [member states] and allies, while maximizing the exploitation of SSA capabilities."

While it is still unclear, which systems and sensors will finally contribute to a European Space Surveillance System, following the current development, it becomes clear that there will always be information gathered by individual sensors, which might be classified or sensitive and thus distributed to other member states only under special conditions. A surveillance system is therefore expected to have a data policy in place, which allows to separate classified from un-classified (orbit) information. As sensor raw data is supposed to represent the best available data, it is thus important to have methods and procedures available allowing the operating entity to forward de-classified and, maybe, also degraded orbit information to their users.

Orbit determination and satellite catalogue maintenance

The sensors within a space surveillance network collect observational data for the orbit determination (OD) process, which aims at providing a set of orbital elements for each tracked object. For the SSN, the publicly available orbital elements are TLE, which contain doubly averaged Keplerian elements (and mean motion instead of semi-major axis). A simplified scheme of the cataloguing process chain, showing an exemplary sensor at its top and the satellite catalogue database as its final product, is shown in Figure 1.4.

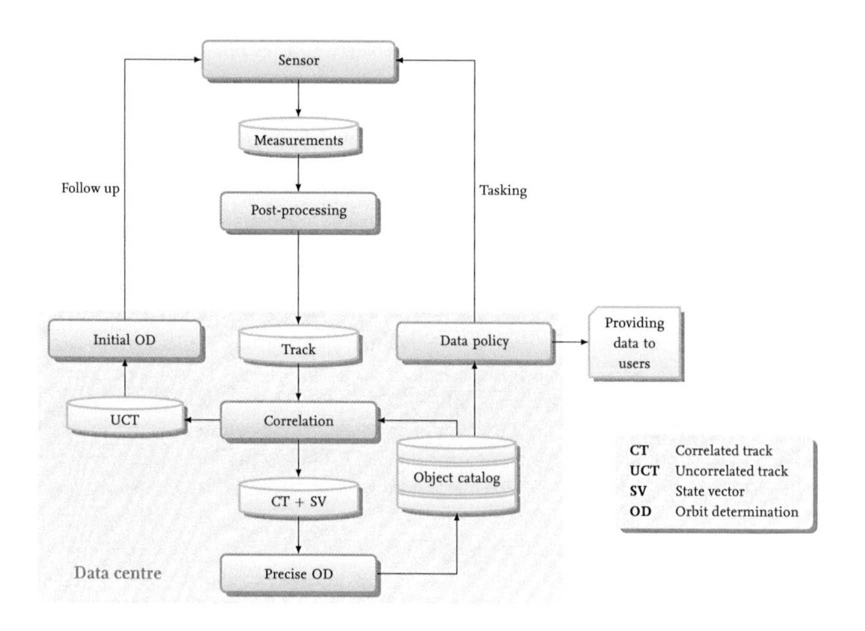

Figure 1.4.: A simplified scheme of the cataloguing process chain. Observational data from a sensor is processed to finally result in a catalogue update. The data is subject to a data policy scheme, which may contain a declassification, a conversion or fit to provide orbital elements, or an ephemeris compression.

The type of measurements obtained depends on the sensor used. Radars would typically provide information on range, range-rate, elevation and azimuth, while telescopes are restricted to angles-only observations, i.e. elevation and azimuth, in general. Individual observations are combined to form a *tracklet*, which contains all observations of a single object obtained during one station pass. Single or multiple tracklets are then correlated with catalogued objects in order to see, whether the observed object matches the predicted trajectory of a catalogued object. If the correlation is successful (denoted as a Correlated Track (CT)) a catalogue update is possible for the newly obtained data. Therefore, the CT as well as the State Vector (SV) (in its simplest representation, a state vector combines the radius and velocity vector for a given epoch, often also referred to as an *ephemeris*) from the catalogue are passed to a process known as *Statistical Orbit Determination* (or *Precise Orbit Determination*), which uses estimation techniques to incorporate the new information into the SV update.

For unknown objects, an Uncorrelated Track (UCT) is passed to the *Initial Orbit Determination* (IOD), which provides a first orbital arc, allowing for follow-up observations and subsequent orbit refinement through statistical orbit determination.

The data centre is responsible for tasking sensors, e.g. for observing objects that require a catalogue update. Finally, data may be provided to users subject to a data policy scheme.

An important design parameter for an SSA system is the *update interval* of the catalogued objects. Due to orbit determination uncertainties and modelling errors in the propagation techniques, the computed trajectory will degrade over time with respect to the true orbit. In order to re-acquire an object and being able to correlate it with its catalogue entry, it is thus necessary to re-observe it. A fundamental quantity to assess when it is required to update the existing orbit information for an object is the *Fisher information* (Frieden, 1998). It measures the available amount of information about the state vector of an object, which is assumed as a random variable. Being a function of the object's orbit, the Fisher information allows to derive how often an object needs to be re-observed. In practice, however, the update interval is shorter, as satellites perform manoeuvres, or more frequent observational information is required for single objects. The latter is typically associated with collision avoidance (CA) operations, where orbit information has to be valid for a time span sufficient to verify, implement, upload and execute a manoeuvre (Krag et al., 2010).

If orbit determination is performed by means of a batch least squares process, an additional quantity, the *fit span*, needs to be defined as the number of observations taken into account for a single orbit update. It is important to note that, while update intervals may be quite short, fit spans used for each update may overlap and are therefore not necessarily linked to the update cycle.

Orbit determination accuracy

The most accurate Orbit Determination (OD) results today are obtained for a very limited set of geodesy and oceanography satellites. The combination of Satellite Laser Ranging (SLR), Doppler Orbitography and Radiopositioning Integrated by Satellite (DORIS) and Global Positioning System (GPS) measurements, as well as a fit span of several days (e.g. 10-day orbital arcs are used for TOPEX/Poseidon (IDS website, 2015)) provides solutions with an accuracy on the order of 1 cm in the radial component[5]. So-called Precision Orbit Ephemerides (POEs) are generated with a further delay of several weeks, in order to incorporate the latest environmental information (IDS website, 2015).

For near real-time data, Klinkrad (2006) provides examples for the OD accuracy of ESA's ERS-2 and Envisat satellites, both were operated in sun-synchronous orbits. The radial position accuracy (standard deviation) is 0.5 m, while the other components are 1.0 m in out-of-plane and 3.0 m in along-track direction (Klinkrad, 2006). The corresponding velocity errors are 1.0 mm/s for out-of-plane and along-track directions and 3.0 mm/s in radial direction (Klinkrad, 2006).

[5] `http://ids-doris.org/organization/about-ids.html`, accessed on January 15, 2015.

While the previous examples are valid for single satellites under certain conditions, specifying the capabilities of a surveillance network is difficult. For a given object of the catalogue, the orbit determination result will be strongly impacted, for example, by the number of observations and dedicated trackings, the orbit itself, or the size and area-to-mass ratio of the object. Rough estimates of the bias and noise characteristics of some SSN sensors are given by Vallado and McClain (2013). For example, the AN/FPS-85 (Eglin) radar is given with a range measurement noise of 32.1 m, while the corresponding values for azimuth and elevation are 55″ and 53″, respectively. While for a single measurement, using these values, a satellite at 1000 km altitude would be associated with an error of about 260 m in azimuth and elevation direction, this information is not comparable to the OD results, which process multiple observations using SP techniques.

Deriving SP catalogue uncertainties from data messages

Taking into account all relevant perturbations, including all secular and periodic contributions, and integrating the equations of motion numerically, SP have become a standard today. USSTRATCOM's SP orbits are classified, but, as a result of the Iridium-Cosmos collision in 2009, in July 2010 USSTRATCOM started sharing so-called CSMs with non-USG entities, especially with satellite O/O. In April 2014, the CSM was replaced by the standardised CDM, its format being defined by the Consultative Committee for Space Data Systems (CCSDS) (CCSDS 508.0-B-1, Blue Book, 2013). With the main intention being to support Collision Avoidance (CA) operations, an operational CDM, as provided by JSpOC provides the variances and covariances of the position vector for a specific conjunction event at the time of closest approach (TCA). Although CDM information contains propagated values for the *covariance matrix*, it is still possible to roughly estimate the order of magnitude of the SP state vector errors at epoch. As an example, average values for the position uncertainties were obtained in a detailed CSM analysis for ESA's Assessment of Risk Event Statistics (ARES) tool and showed large variations for different orbits and object sizes (Sánchez-Ortiz et al., 2013). The results range from large objects (RCS>0.1 m^2) with 1-σ uncertainties in the order of magnitude of 10 m to errors for small objects in the km-regime (Sánchez-Ortiz et al., 2013).

Uncertainties in the TLE data

The public subset of the TLE catalogue as provided by USSTRATCOM, results from an analytical (or General Perturbations (GP)) theory, which was pioneered by Brouwer and later adapted by Lane, Cranford and Hujsak (Hoots and Roehrich, 1980). The currently used models are Simplified General Perturbations (SGP4) for all orbits with a period of $T < 225$ min, and Simplified Deep-space Perturbations (SDP4) for deep-space objects with $T \geq 225$ min (Hoots and Roehrich, 1980).

Until 2013, the SGP4/SDP4 models were directly applied to the observations, in parallel with the SP techniques, to obtain an orbit fit over the observation span. The mean Keplerian elements were found for a TLE epoch which was close to the latest observation and typically near the ascending node of the orbit.

Table 1.1.: Averaged results for component uncertainties (in metres) in satellite-centered UVW frame from orbit determination with TLE as pseudo-observations based on catalogue snapshot from 2008-01-01. (Flohrer et al., 2008)

Perigee altitude / km	Direction	Standard deviation / m					
		$e < 0.1$			$e > 0.1$		
		i / deg			i / deg		
		< 30	30 to 60	> 60	< 30	30 to 60	> 60
< 800	Radial (U)	67	107	115	2252	629	494
	Along-track (V)	118	308	517	4270	909	814
	Cross-track (W)	75	169	137	1421	2057	1337
800 to 25 000	Radial (U)	191	71	91	1748	1832	529
	Along-track (V)	256	228	428	3119	1878	817
	Cross-track (W)	203	95	114	971	1454	1570
> 25 000	Radial (U)	357	-	-	402	4712	-
	Along-track (V)	432	-	-	418	6223	-
	Cross-track (W)	83	-	-	83	1208	-

According to Bowman, 2014, that process changed in the early months of 2013: For a "great majority of objects" (Bowman, 2014), SP were now used to first obtain an accurate ephemeris for the fit span. The next step is to perform a 3-day prediction of the ephemeris, where also solar indices and the Disturbance storm time (Dst) index of the JB2008 model are forecast. The extrapolated ephemerides are then used to generate a fit with the SGP4/SDP4 model to provide the TLE, with the TLE epoch now being at the beginning of the predicted ephemerides.

From an operational point of view, the new procedure allows to maintain only one SP catalogue, while GP-based TLE data sets may be generated on demand from SP ephemerides and do not require their own catalogue, which was already outlined by Schumacher and Hoots (2000) and Wilkins et al. (2000) shortly after SP techniques started being integrated into routine operations.

Recalling the motivation for the development of the SSN, the TLE data format was originally intended for tracking purposes. Being the only comprehensive space object data source, however, it is widely used for all kind of applications, also in conjunction analysis. In the latter case, covariance information is required, but it is unavailable in the TLE format. Thus, several studies in the past have focused on estimating uncertainties in the TLE data, e.g. Kelso (2007), Flohrer et al. (2008), Levit and Marshall (2011), Aida and Kirschner (2011) or Kahr et al. (2013). As an example, the average results for different orbit classes from Flohrer et al. (2008) are shown in Table 1.1.

The results in Table 1.1 were obtained by generating pseudo-observations using TLE and the SGP4/SDP4 theory. The orbit determination, based on SP techniques, then provided a fit for an orbital arc of 24 h with the residuals (as shown in Table 1.1) representing the combined errors of both, the analytical theory behind SGP4/SDP4 and the numerical theory used for the fit (Flohrer et al., 2008). A general difficulty using such an approach is to select a suitable weighting matrix (see also Section 3.4.1). In Flohrer et al. (2008) a diagonal weight matrix was used, where the three radius components had equal weights. The velocity components weights were scaled by 10^{-3} compared with the radius components and were also of equal size.

It can be seen in Table 1.1 that for low-eccentricity ($e<0.1$) orbits, the radial and cross-track position errors in most cases are in the order of magnitude of 100 m, while the along-track component may be in error by up to about 500 m. For high-eccentricity orbits ($e>0.1$) the obtained errors were up to a few kilometres in some cases.

The above procedure assumes that TLE are unbiased and individual objects are consistently tracked, which might not be true (Vallado and Cefola, 2012). In fact, the new procedure to generate TLE data, as employed by USSTRATCOM in 2013, was already analysed by Wilkins et al. (2000) and the results showed a major improvement in accuracy for the subset of SLR satellites analysed in that work. For high-altitude objects the accuracy of 1 km to 5 km, for a propagation of a few days, was reduced by a factor of 5 to 10 (Wilkins et al., 2000). The GP theory thus seems to provide better results, after the raw observations have been already smoothed by SP techniques (Wilkins et al., 2000). Also, as the SGP4/SDP4 model is used to generate a fit on a forecast trajectory with numerical accuracy, the propagation error of the analytical theory can be reduced (Wilkins et al., 2000).

Identification of orbital data users and associated requirements

In contrast to the US and Russian surveillance system, the initial design for the European SSA system is based on the requirements of a wide range of different users (or customers), also because SSA is defined in a much broader context as stated above. According to Bobrinsky (2009), the following entities could be included:

- European governments (EU, national, regional)
- European space agencies
- Spacecraft operators (commercial, academic and governmental)
- Academic and research institutions
- Space insurance and space industry
- Energy industry, including surveying, electricity grid operators, electrical power suppliers
- Network, telecommunication and radar system operators
- Space weather service providers
- European and other air traffic control and navigation service providers
- European and international rescue and disaster-response authorities
- United Nations and other international bodies

- Defence sector / defence security

Of course, the identified users and user groups are in need of different information and rely on several distinct services the SSA system has to provide. A general overview on the basic products, services and user groups for the Space Surveillance and Tracking (SST) segment of the SSA system is provided in Figure 1.5, based on Krag et al. (2010).

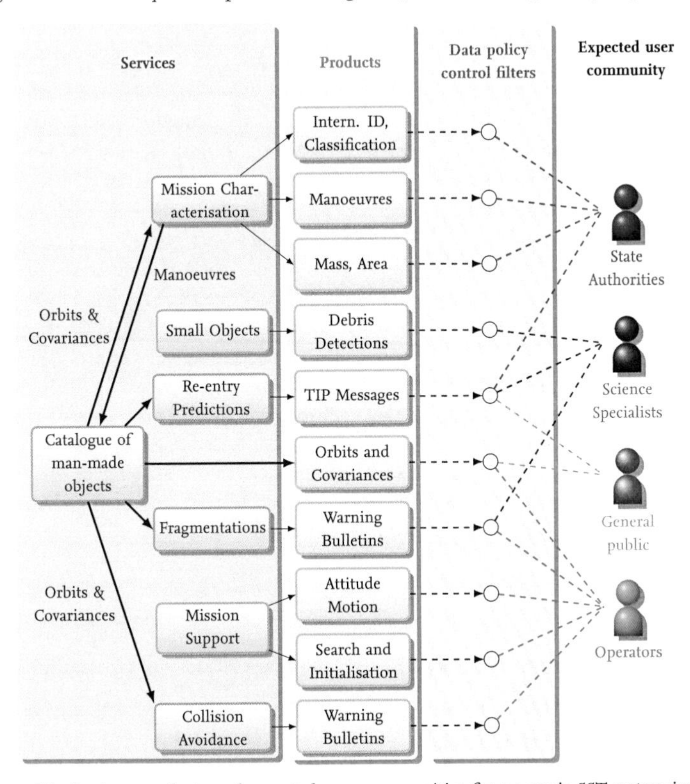

Figure 1.5.: Services, products and expected user communities for a generic SST system including data policy control filters. Note that red-colored boxes represent segment services which are provided as required. (according to Krag et al. (2010)).

As can be seen, a catalogue of on-orbit objects is the basis of most services and the resulting products, for example the Tracking and Impact Prediction (TIP) message specifically used in the Re-entry service domain. In the context of this thesis, the provision of orbits and covariances shall be focused on, which are associated with two expected communities (general public and operators) in Figure 1.5 through dedicated data policy control filters. In addition, orbits and covariances are also accessed indirectly via the *Mission Characteri-*

sation and *Collision Avoidance* services. The data policy control filters are the top layer, as seen by the users, and shall fulfill several tasks. First of all, following the idea of combining available national assets (sensors) into a European system, the design should incorporate national security requirements. These might restrict users from exchanging sensitive information across national borders, which would be the case if individual sensors are also used in the defence branch and observations or processed orbit data is only allowed to be forwarded with a degraded accuracy. One very popular example might be the past use of Selective Availability (SA) for the GPS satellites being owned and operated by the US government, where signals were deliberately degraded resulting in a limited position accuracy significantly below the system capabilities. Also, the examples of the Russian (not accessible to the public) and the US system (SP data not accessible, except for CDMs) show that there is an incentive to restrict a full catalogue access.

From a commercial point of view, it might be advantageous to provide orbits and covariances in different accuracy regimes. Users in need of high-accuracy trajectories would pay a different price than those who can afford using orbits with less accuracy within their applications.

The above mentioned concept is similar to the one applied by USSTRATCOM, where SP ephemerides are for military users only, while less accurate TLE are provided to the general public. However, especially for CA operations, missing information on the uncertainty in the provided TLE data poses some problems, as illustrated in an Envisat example by Krag et al. (2010): On January 21, 2010, Envisat was supposed to have a close encounter with a 3.8 t upper stage, where the miss distance was only 48 m, but analysis of the TLE data showed a radial separation of 346 m. Only additional tracking, using the TIRA radar, revealed a radial separation of only 15 m, which was confirmed by the CSM information and ultimately resulted in an avoidance manoeuvre.

Based on the user requirements, the critical design and cost drivers can be analysed, where the following were identified by Krag et al. (2010):

- The lower diameter cut-off above which catalogue coverage has to be provided for a defined coverage level,
- the accuracy of the orbit information provided and
- the overall technical availability of the system.

As the focus of this thesis shall be on orbit information accuracy, the second of the above requirements shall be explained in more detail in the following.

The overall achievable accuracy for any on-orbit object is limited by the quantity and quality of the observations obtained. Especially the use of SP over GP techniques provides for a shift from "a model-limited system to a data-limited system" (Schumacher and Hoots, 2000). With SP models having an inherent error on the order of magnitude of a few metres (Schumacher and Hoots, 2000), this basically sets the level for the maximum achievable accuracy of a system, although there are still deviations to be expected for varying object size, high-eccentricity orbits or even the small subset of satellites equipped with sensors to provide solutions in the cm-regime using SLR, GPS and DORIS.

While these numbers are valid for the orbit state at or close to the orbit determination epoch, the accuracy tends to degrade over time when the state is being propagated. For an operational system it is therefore important to define an update cycle, which makes sure that the propagation remains within a pre-defined accuracy envelope.

For an SP catalogue, the accuracy envelope can be evaluated by referencing the covariance information. According to Krag et al. (2010), collision avoidance appears to be the most demanding service regarding the definition of an accuracy envelope. With the target satellite orbit determined with an accuracy of 1 m to 10 m (Klinkrad, 2006; Krag et al., 2010), the orbits of the chaser objects should be provided in a comparable accuracy regime (Krag et al., 2010).

1.3. Research approach

1.3.1. Objectives

The European SSA system is currently being developed following an approach oriented towards a broad range of different users. Design criteria are obtained from the expected services the system is intended to provide (Figure 1.5). For the SST segment, it has been identified that CA operations pose the most demanding requirements regarding the accuracy of the orbit data in the catalogue, both for the state at orbit determination epoch as well as for the required update intervals due to accuracy degradation, when states are propagated.

The cataloguing based on SP models, as described for the US system, provides highly accurate solutions, while TLE data, as a result from an analytical (GP) model, are distributed to the public for all kinds of applications. While the US approach shows that there seems to be a motivation in providing different users with orbital information of varying accuracy, using TLE is associated with some shortcomings. The most significant one is that TLE come without any metric on the inherent error. Several authors have thus suggested that a catalogue of orbital objects ideally should provide information on the accuracy associated with the individual products: according to Vallado and McClain (2013), a surveillance system could provide data "with an accuracy the user requires".

This has several advantages: First, for a commercial system, a *product differentiation* allows for a different pricing according to the quality of the provided data. Although the data provided by JSpOC is freely available at the moment, a possible commercialization does not seem unreasonable in view of the recent contract awarded by the Pentagon to Analytical Graphics Inc. (AGI) for the provision of orbit data through its Commercial Space Operations Center (ComSpOC) service augmenting the existing JSpOC (SpaceNews, 2015).

On the other hand, even without a commercial idea in mind, a product differentiation can be desirable. Providing highly accurate ephemerides and associated uncertainties implies a significantly increased amount of data to be transferred and stored, as opposed to using TLE for tracking purposes only. For example, Oltrogge and Kelso (2011) analysed that ephemerides in LEO have to be provided in about three minute steps to preserve a 50 m position accuracy level via interpolation.

The first objective or scientific question of this thesis shall thus be defined as:

(1) Is it possible to derive, from available high-accuracy data, a solution with a pre-defined accuracy tailored to the user's needs?

An important tool to perform such an analysis is an orbit propagator based on numerical integration. While there are many such tools available, only very few have been designed specifically with an SST context in mind. Of those, only one or two are known to the author of including the assessment of process noise in the propagation of the covariance matrix. Moreover, access to the source code and manipulations thereof were required for the specific analyses to meet objective (1).

(2) Design an orbit propagation tool based on numerical integration for state vector and covariance extrapolation, as well as the means to assess process noise. This propagation tool shall be specifically designed to be used in an SST system.

With a method to provide orbits with pre-defined accuracy at hand, the result will be ephemerides that can be provided to the users via standardised orbit data messages, like the Orbit Ephemeris Message (OEM) or the Orbit Parameter Message (OPM) (CCSDS 502.0-B-2, 2009). However, this means that the user still has to take care of interpolating in between the data points, likely to result in a degradation of the accuracy present in the delivered data message. It is thus desirable for the user to extract orbit information from the data message at any point in time without introducing errors. This leads to the next objective:

(3) Provide the means to establish a data message containing continuous information making use of current data message standards.

While standardised messages like the OPM define the structure to provide information on the orbit uncertainty in terms of the covariance matrix, this information typically is either distributed in an incomplete state (e.g. only a subset of the full matrix as was the case for the CDM provided by JSpOC until January 19, 2016) or completely missing. The covariance matrix, which is a direct result of the orbit determination process is part of the SP catalogue.

(4) Investigate a method to also provide interpolated covariance matrix information associated with the obtained state vectors of predetermined accuracy.

1.3.2. Methodology

The defined objectives required to study several interdisciplinary topics, which are outlined in a mindmap shown in Figure 1.6. The theoretical background of many of those topics needs to be adressed. It was decided to combine the background with the obtained results, where appropriate, as opposed to have the full theoretical background first and

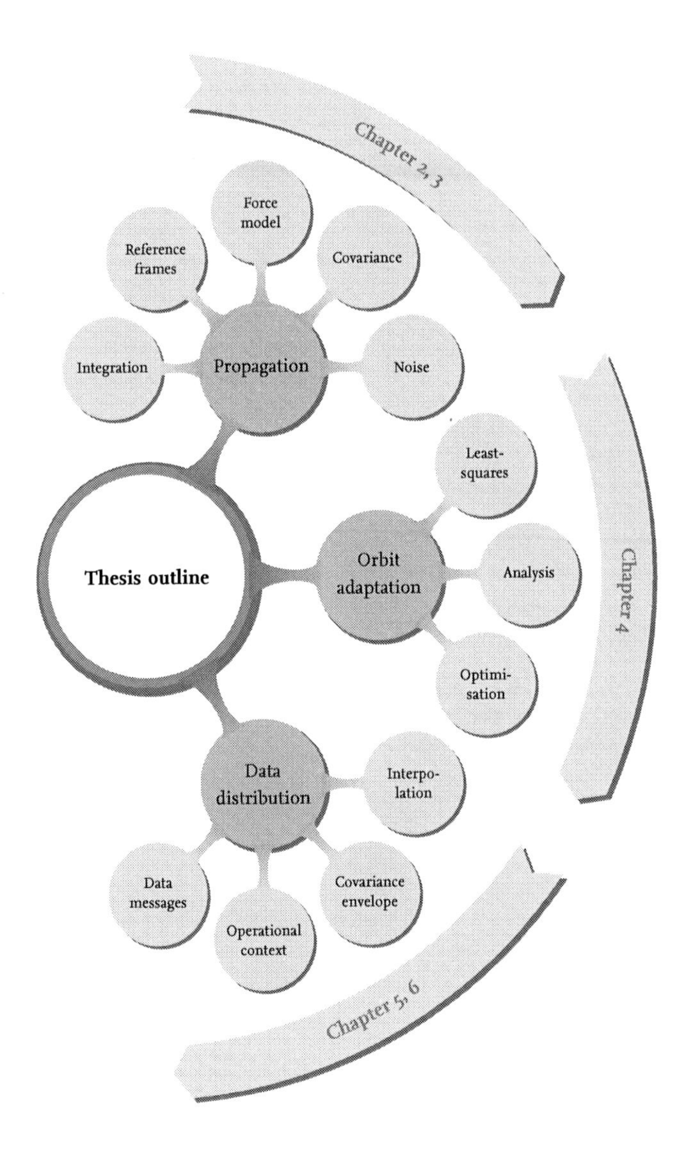

Figure 1.6.: Mindmap showing the different relevant topics to the study objectives, as well as the associated chapters in this thesis.

the findings in later chapters. This was considered to be less confusing to the reader, in view of the sometimes non-related contents, while at the same time being a more enjoyable read to those who already have some background on the subject. One example is the introduction to process noise resulting from the geopotential series truncation and coefficients uncertainties, which is presented together with exemplary results.

The numerical propagation tool NEPTUNE , introduced in Chapter 2, was carefully designed focusing on its potential use within a space surveillance system and the associated cataloguing process. The force model was selected to match current orbit determination capabilities, but also to be in line with current standards and guidelines for orbit propagation. The integration of both, the state vector and the covariance matrix, was designed to meet performance requirements for a catalogue of several thousand objects. In addition, the assessment of the cross- and auto-correlation of the process noise, was included to allow for a more realistic prediction of the covariance matrix - which is essential for the services an SST system provides.

The next step was to design and implement an algorithm which allows to derive orbits of predetermined accuracy with respect to a reference trajectory generated by the Networking/Partnering Initiative (ESA research programme with industry and academia) (NPI) Ephemeris Propagation Tool with Uncertainty Extrapolation (NEPTUNE)[6].

This approach is justified by the fact that even in an operational system, raw orbit measurements are likely to be processed first by a numerical theory, and only the resulting smoothed results are then used in subsequent processing.

The method to generate orbits with predetermined accuracy is based on a modification of the geopotential perturbations and will be referred to as Geopotential Adaptation Method to Bias Trajectories (GAMBIT) in the following.

In order to provide the GAMBIT-derived data with the associated covariance matrix, interpolation techniques were developed (Chapter 5). These allow for continuous orbit information for a given time interval within a data message. The interpolation of a computed envelope function for the variances of the state vector results in a significant compression ratio.

The implications for the operational context are analysed and discussed in Chapter 6, bringing all the developed methods together and showing how parts of an SSA system data policy layer containing the ideas from this thesis might look like.

[6]The Networking/Partnering Initiative by ESA aims at supporting the development of advanced space-related technologies in research institutes and universities and strengthen the links between ESA and European institutions (http://www.esa.int/Our_Activities/Space_Engineering_Technology/Networking_Partnering_Initiative, accessed on October 17, 2016).

2

Orbit dynamics and force modelling

One of the key elements of a space surveillance system is the orbit propagation technique employed in the orbit determination process. While the first surveillance systems in the US and the Soviet Union were using fast analytical techniques, also known as GP methods, with today's computational limitations virtually non-existing, it is common to use so-called SP techniques, which perform the numerical integration of an arbitrarily detailed force model.

In this chapter, the force model of the numerical propagation tool NEPTUNE (Networking/Partnering Initiative (ESA research programme with industry and academia) (NPI) Ephemeris Propagation Tool with Uncertainty Extrapolation) is described, which was developed as the important core element for the subsequent analyses in this dissertation. The major requirement in the design of NEPTUNE was to focus on its use within a space

Figure 2.1.: NEPTUNE logo.

surveillance context, while at the same time being compatible with prevailing international standards and recommended practices, like ISO 11233:2014 (ISO, 2014) or ANSI/AIAA S-131-2010 (ANSI/AIAA, 2010).

The force model has been selected to cover all major perturbations that allow to obtain orbit solutions in the 10 m regime, as outlined in Section 1.2 for the space surveillance context. However, it is not a trivial task to decide on which perturbations are relevant and which ones can be neglected. Especially, if the tool is going to be applied to any object and orbital region, even small perturbations may result in a secular drift and thus become important due to an accumulation of errors. The selected force model represents a trade-off, as many perturbations require very complex modelling. For example, the Yarkovsky thermal radiation requires detailed knowledge about the material properties of the object, heat conduction through the structure, and its attitude motion. Nevertheless, the NEPTUNE software was designed in a very modular way, so that additional perturbations can be conveniently included.

As Neptune was required to be applicable to any Earth orbit, another important step is to select an appropriate numerical integration technique. While a fixed step integration works seamlessly for near-circular orbits, using a varying-step method for high-eccentricity orbits is beneficial.

The following list gives an overview on the characteristics of Neptune , while the subsequent sections will provide more details:

- State vector integration: variable-step, multi-step, double-integration in Predict-Evaluate-Correct (PEC) mode.
- Covariance matrix integration: Formulation using variational equations, with RK4 integration of the state error transition matrix.
- Non-spherical geopotential: Three different models are supported: EIGEN-GL04C, EGM96 and EGM2008.
- Atmospheric drag: Density via the NRLMSISE-00 model and horizontal winds by the HWM07 model.
- Gravitational perturbations by the Sun and the Moon using DE-405 and -421 ephemerides.
- Solar Radiation Pressure (SRP) with a conical Earth shadow including umbra and penumbra regions based on a spherical Earth.
- Earth radiation pressure in the visual (albedo) and infrared regime.
- Solid Earth and ocean tides.
- Reference frames: GCRF/ITRF according to the IAU 2006/2000A theory.

Besides a 3-Degrees-of-Freedom (DOF) numerical propagation for the state vector, Neptune has also the capability to propagate the covariance matrix, which is an essential step in the filtering process within the statistical orbit determination, but is also beneficial for the GAMBIT method presented in Chapter 4. An introduction into the implemented models and methods for Neptune is provided in the following sections. More detailed information, especially related to the implementation with respect to accepted best practices, is given in Annex A.

2.1. Numerical integration of the equations of motion

The equations of motion are defined in an inertial cartesian reference frame. Using Cowell's formulation (Cowell and Crommelin, 1909) to combine the perturbations with Newton's gravitational law, they can be written as:

$$\ddot{\mathbf{r}} = -\frac{\mu}{r^2}\frac{\mathbf{r}}{r} + \mathbf{a}_p, \tag{2.1}$$

here, $\mathbf{a}_p$ is a superposition of the accelerations acting on the satellite due to the modelled perturbations. The formulation in Equation 2.1 allows for a simple numerical integration, a technique referred to as Special Perturbations.

A variable-step double-integration multi-step Störmer-Cowell method (Berry, 2004) was selected, which is appropriate for space surveillance applications: as Berry (2004) points out, one of the main criteria for the selection of the integrator is the number of force function evaluations.

With the force model being very complex, the integrator should tend to minimise the number of evaluations. At the same time, the overhead of the integrator itself is negligible. In fact, for NEPTUNE the force model function evaluation takes up more than 90 % of the total runtime. Furthermore, as the method is not variable order and double-integration[1], the additional stability gained thereby allows to have a minimum number of function evaluations per step: the integrator follows a PEC scheme, with only one evaluation per step.

A variable stepsize is advantageous, especially for eccentric orbits, where stepsize control with pre-defined tolerances allows to take larger steps when the object is at apogee.

A brief overview on the integrator is given in the following sections. For a detailed description of the method, refer to Berry (2004).

2.1.1. Prediction

The first step in the integration is the prediction of a new position vector $\mathbf{r}^p_{n+1}$ based on a set of k backpoints. The *general formulation* predictor is given by Lundberg (1981) for a k-th degree polynomial $\mathbf{P}_{k,n}(t)$ interpolating the backpoints, which is integrated twice in time:

$$\mathbf{r}^p_{n+1} = \mathbf{r}_n + h_{n+1}\dot{\mathbf{r}}_n + \int_{t_n}^{t_{n+1}} \int_{t_n}^{\bar{t}} \mathbf{P}_{k,n}(t)\, dt d\bar{t}, \tag{2.2}$$

with h_{n+1} being the current stepsize. Introducing modified divided differences for the polynomial, Berry (2004) is able to compute the double integral and derives the following formulation for the predictor of the variable-step Störmer-Cowell integrator:

$$\mathbf{r}^p_{n+1} = \left(1 + \frac{h_{n+1}}{h_n}\right)\mathbf{r}_n - \frac{h_{n+1}}{h_n}\mathbf{r}_{n-1} + h^2_{n+1}\sum_{i=1}^{k}\left(g_{i,2} + \frac{h_{n+1}}{h_n}g'_{i,2}\right)\boldsymbol{\phi}^*_i(n), \tag{2.3}$$

here, $\boldsymbol{\phi}^*_i(n)$ are functions of the modified divided back differences and the stepsizes for the individual steps between the backpoints, while $g_{i,2}$ and $g'_{i,2}$ are stepsize-dependent coefficients.

A second integration is required to obtain the velocity vector. The Shampine-Gordon predictor (Shampine and Gordon, 1975) is used:

$$\dot{\mathbf{r}}^p_{n+1} = \dot{\mathbf{r}}_n + h_{n+1}\sum_{i=1}^{k} g_{i,1}\boldsymbol{\phi}^*_i(n). \tag{2.4}$$

Although one could use this predictor to obtain both radius and velocity vectors by subsequent single-integration, this would introduce additional round-off errors, which are minimized by the double-integration approach.

[1]The radius vector is obtained directly upon double integration of the acceleration.

2.1.2. Correction

After having obtained a predicted position vector $\mathbf{r}^p_{n+1}$, a new function evaluation is performed at that point, resulting in a new acceleration. Using that value together with the set of k backpoints, a new interpolation polynomial of degree k can be used for the corrector formula, which is the final step in the PEC cycle.

The variable-step Cowell predictor is:

$$\mathbf{r}_{n+1} = \mathbf{r}^p_{n+1} + h^2_{n+1}\left(g_{k+1,2} + \frac{h_{n+1}}{h_n}g'_{k+1,2}\right)\boldsymbol{\phi}^p_{k+1}(n+1). \tag{2.5}$$

Similarly, the Shampine-Gordon corrector providing the velocity vector is obtained from (see also Annex B.1.3):

$$\dot{\mathbf{r}}_{n+1} = \dot{\mathbf{r}}^p_{n+1} + h_{n+1}g_{k+1,1}\boldsymbol{\phi}^p_{k+1}(n+1). \tag{2.6}$$

2.1.3. Stepsize control

At each step, the local error is kept below a user-defined tolerance, which is a combination of an absolute tolerance, ϵ_{abs}, and a relative tolerance, ϵ_{rel}. The local error for single (superscript s) and double integration (superscript d) is computed as the difference between the corrector results (Equation 2.6 and Equation 2.5, k^{th} degree polynomial) and a corrector using the set of $n+1$ points interpolated by a polynomial of degree k for the velocity and position vectors, respectively.

$$\epsilon^s_l = \dot{\mathbf{r}}_{n+1} - \dot{\mathbf{r}}_{n+1}(k), \tag{2.7}$$

$$\epsilon^d_l = \mathbf{r}_{n+1} - \mathbf{r}_{n+1}(k). \tag{2.8}$$

The components of the local error are then combined in a weighted sum of squares at each step and compared to the tolerance ϵ_{max}:

$$\sqrt{\sum_{i=1}^{3}\left(\frac{\epsilon^s_{l,i}}{w^s_i}\right)^2} \leqslant \epsilon_{max}, \tag{2.9}$$

with

$$\epsilon_{max} = \max(\epsilon_{rel}, \epsilon_{abs}), \tag{2.10}$$

and the weight functions

$$w^s_i = |\dot{r}_i|\frac{\epsilon_{rel}}{\epsilon_{max}} + \frac{\epsilon_{abs}}{\epsilon_{max}}. \tag{2.11}$$

The above equations are similar for double integration (see Annex B.1.4), containing the position instead of the velocity vector components.

The step is successful, if Equation 2.9 is fulfilled for both the single and double integration formulation. Otherwise, the step is repeated with half the stepsize. If this happens three times in succession, the integration is reset and starts as a first-order method again (see also Section 2.1.4). In general, such a restart will be required at discontinuities like shadow boundary crossings, manoeuvres, etc.

After a step was successful, the stepsize for the next step is computed to keep the local error as close as possible to the tolerance. Using $h_{n+2} = \rho h_{n+1}$, assuming that the divided differences are slowly varying and that all preceding steps were taken with h_{n+2}, Shampine and Gordon (1975) derive an equation for the stepsize factor ρ for single integration. Similarly, Berry (2004) provides an analogous formulation for double integration:

$$\rho^{s/d} = \left(\frac{0.5\epsilon_{max}}{\zeta_l^{s/d}} \right)^{\frac{1}{k+1}} \tag{2.12}$$

The approximated local error $\zeta_l^{s/d}$ is the error, that would be made if the previous steps had been taken with h_{n+1}. It is computed for both, single and double integration and the smaller of both values is selected. See Annex B.1.4 for more details.

The calculated value of ρ is bounded between 0.5 and 2.0, so that the stepsize is doubled for all values $\rho \geqslant 2.0$ and halved for $\rho \leqslant 0.5$. While Shampine and Gordon (1975) designed their method in a way preferring constant stepsizes, thereby reducing the overhead for the re-computation of the integrator coefficients, Berry (2004) points out that due to the very expensive force model, the additional integrator overhead can be neglected in favour of having a fast increase of the stepsize towards larger values. Therefore, Berry (2004) recommended to have the same boundaries at $\rho = 0.5$ and $\rho = 2.0$, without any further restrictions for values in between.

2.1.4. Initialization

An important aspect for multi-step integration is the initialization or startup procedure. With only the initial state vector being available at t_0, either a set of backpoints have to be computed before starting the multi-step integration, or the integrator has to start as a variable order method. For the former approach, it is possible to use a single-step integrator, or even a Taylor series representation, until the required number of backpoints are found and then switch to the multi-step integration. The variable order startup is designed to start as a first-order method with small stepsizes and both the order and the stepsize are increased in subsequent steps.

The Störmer-Cowell method used for NEPTUNE , is based on a variable-order startup (Berry, 2004). The initial stepsize is found by estimating the local error of a first-order method (Berry, 2004):

$$h \approx \sqrt{\frac{\epsilon_{max}}{|\ddot{\mathbf{r}}_0|}} \tag{2.13}$$

For a more conservative estimate, that value is divided by four and the individual components of the acceleration vector are weighted:

$$h = \frac{1}{4} \sqrt{\frac{\epsilon_{max}}{\sqrt{\sum_{i=1}^{3} \left(\frac{\ddot{r}_{0,i}}{w_i^s} \right)^2}}}. \tag{2.14}$$

An upper boundary equal to the first requested output is set on h, in case the accelerations at the initial point are close to zero. A lower boundary to reduce round-off error is also required and set to $4\epsilon_m t_0$, where ϵ_m is the *machine epsilon.*

While the initial step size in Equation 2.14 is computed for single integration, the double integration formulation of Equation 2.14 is analogous:

$$h = \frac{1}{4}\sqrt{\frac{\epsilon_{max}}{\sqrt{\sum_{i=1}^{3}\left(\frac{\dot{\mathbf{r}}_{0,i}}{w_i^d}\right)^2}}}, \quad (2.15)$$

using the velocity vector instead of the acceleration and the double-integration weight functions associated with the velocity. The Störmer-Cowell integrator then uses the smaller of both values in order to stay within the requested tolerance.

The first order predictor then is:

$$\mathbf{r}_1^p = \mathbf{r}_0 + h\dot{\mathbf{r}}_0 + \frac{1}{2}h^2\ddot{\mathbf{r}}_0, \quad (2.16)$$

and the corrector:

$$\mathbf{r}_1 = \mathbf{r}_1^p + \frac{1}{6}h^2\boldsymbol{\phi}_2(1). \quad (2.17)$$

In order to have a faster startup, the initialization will only double (halve) the stepsize, if a step succeeds (fails). For additional stability, a second function evaluation is introduced, resulting in a PECE cycle. After the first successful step, the variable-step Störmer-Cowell integrator is started, beginning with a first-order polynomial, increasing the order after each successful step until a set of nine backpoints is available and the method continues as an eighth-order polynomial without further changing the order.

2.1.5. Interpolation

In principle, the integrator can be configured to have a stepsize corresponding to the requested output time. As this implies that the stepsize could be smaller than the one obtained from the stepsize control according to the error tolerance, an unnecessary increase in computation cost would be introduced. Therefore, Shampine and Gordon (1975) give an interpolation formula to find the output value at the requested time t_I, which is between the points n and $n+1$. Using the $(k+1)^{th}$ degree polynomial for the set of backpoints, the interpolated value is found via (Shampine and Gordon, 1975):

$$\dot{\mathbf{r}}_I = \dot{\mathbf{r}}_{n+1} + \int_{t_{n+1}}^{t_I} \mathbf{P}_{k+1,n+1}(t)\,dt \quad (2.18)$$

Performing a similar derivation as for the predictor (Annex B.1.2), one obtains for the single integration:

$$\dot{\mathbf{r}}_I = \dot{\mathbf{r}}_{n+1} + h_I\sum_{i=1}^{k+1} g_{i,1}^I\boldsymbol{\phi}_i(n+1), \quad (2.19)$$

with the interpolation stepsize $h_I = t_I - t_{n+1}$ and another set of coefficients, $g_{i,1}^I$, which need to be computed for the interpolation, see also Annex B.1.5.

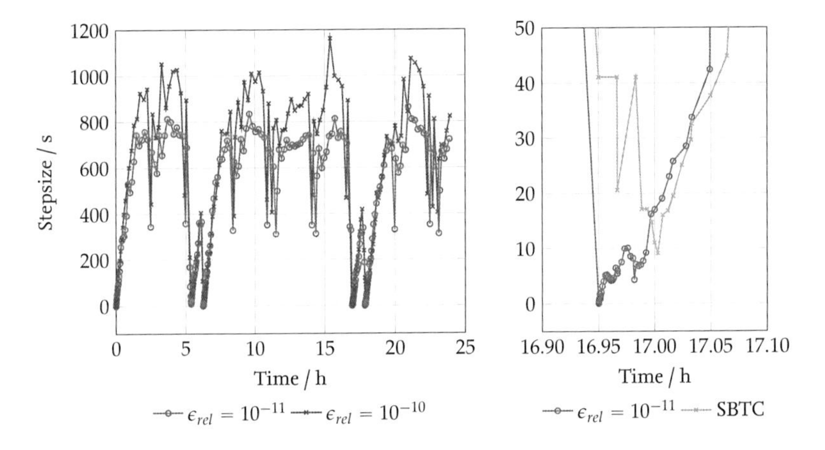

Figure 2.2.: Stepsize for a 24 h propagation of a MEO navigation satellite. Force model: 12×12 geopotential, drag, SRP, luni-solar gravity. Left: Two different levels for the error tolerance are shown, $\epsilon_{rel}/\epsilon_{abs} = 10$. Right: Comparison showing the Shadow Boundary Transit Correction (SBTC).

Analogously, Berry (2004) derives the interpolation formula for the double integration:

$$\mathbf{r}_I = \left(1 + \frac{h_I}{h_{n+1}}\right) \mathbf{r}_{n+1} - \frac{h_I}{h_{n+1}} \mathbf{r}_n + h_I^2 \sum_{i=1}^{k+1} \left(g_{i,2}^I + \frac{h_I}{h_{n+1}} g_{i,2}^{I'} \right) \boldsymbol{\phi}_i(n+1). \tag{2.20}$$

2.1.6. Integrator optimization

The numerical integration presumes continuous and sufficiently smooth functions for the accelerations obtained from the force model. However, this cannot always be guaranteed for the propagation of Earth orbits. For instance, orbital manoeuvres introduce a step change in the acceleration at manoeuvre start and end. Another example is the re-entry phase of a satellite, where elements like solar panels might break off due to an excessive drag load, again resulting in a step increase for the acceleration of the parent spacecraft, as parameters like mass and cross-section change instantaneously.

Besides those special cases, there are orbit perturbations that cause a step increase in the acceleration with periods on the order of the orbital period. The major contribution, especially for high-altitude orbits, is due to SRP, where shadow boundary transits cause either a step in the second (for umbra-only models) or the third derivative (for umbra-penumbra models), respectively. This is exemplarily shown for a 12 h Medium Earth Orbit (MEO) in Figure 2.2, which also illustrates the stepsize control of the integrator. In the left figure, two examples are shown for a relative tolerance of $\epsilon_{rel} = 10^{-10}$ and $\epsilon_{rel} = 10^{-11}$, respectively, with $\epsilon_{rel}/\epsilon_{abs} = 10$ in both cases. The 24 h propagation ($\approx$ two orbital revolutions) shows the initialization phase in the beginning, starting with a variable-order integrator and small stepsizes, followed by an increase in the stepsize to above 10 min in both cases. The shadow phases for the two revolutions can be clearly seen by the pair of integrator resets at the shadow entry and exit (shortly after 5 h and at about 17 h into the propagation). Moreover, for $\epsilon_{rel} = 10^{-10}$, the integrator accepts larger stepsizes (which is expected) and, while the stepsize decreases significantly at the shadow boundary transits, there are fewer resets ($h \approx 0$ s) when compared with the $\epsilon_{rel} = 10^{-11}$ case.

An integrator reset, which is also shown in a close-up for the second shadow entry in Figure 2.2 on the right plot, implies a significant amount of force model evaluations for the start-up and the subsequent stepsize increase. In the example, there are 37 steps, until the stepsize again takes on values greater than 10 s. The adverse effect on the integrator performance is not only the additional required runtime, but also additionally accumulated round-off error.

A method for fixed-step integration was proposed by Lundberg et al. (1991), referred to as Shadow Boundary Transit Correction (SBTC) hereafter, which, upon detecting a shadow boundary transition, performs an update on the set of backpoints: for a shadow entry, the acceleration due to SRP is subtracted from the backpoints, while for the shadow exit they are added to the backpoints, thus guaranteeing a smooth interpolation polynomial.

Horstmann et al. (2015) applied this algorithm to the variable-step Störmer-Cowell integrator. An example is shown in Figure 2.2 on the right, where the correction (using the same error tolerance) does result in a significant stepsize decrease, but a reset can be avoided. Furthermore, the number of steps for the time interval shown in Figure 2.2 is distinctively lower when compared with the uncorrected integration. Horstmann et al. (2015) estimated a run-time performance gain of a few percent for high-altitude orbits using the SBTC.

2.2. Coordinate and time systems

In orbital mechanics, the use of different coordinate and time systems is unavoidable, which is mainly because the equations of motion are most conveniently solved in an inertial frame, while force model evaluations, for example for the geopotential or drag perturbations, are relative to a co-rotating Earth-fixed frame.

Moreover, with many different entities and organizations being involved in SST applications and related services, the compliance with international standards for data exchange and associated reference frames is compulsory.

Based on the resolutions of the International Astronomical Union (IAU) and International Union of Geodesy and Geophysics (IUGG) the International Earth Rotation and Reference Systems Service (IERS) realised and defined a celestial and a terrestrial reference frame (Luzum and Petit, 2010). It is important to differentiate between *reference systems* and *reference frames*: While a reference system is the conceptual definition of a coordinate system, a reference frame is the actual realization, using observations, station coordinates, etc. (Seidelmann, 2006).

The International Celestial Reference System (ICRS) was defined by the IAU Resolution A4 (1991) and was refined in 2000 and 2009 (Luzum and Petit, 2010). For Earth applications, a geocentric reference system is advantageous, with the Geocentric Celestial Reference System (GCRS) being the Earth-centered counterpart of the ICRS (Luzum and Petit, 2010).

The International Terrestrial Reference System (ITRS) is based on the IUGG Resolution 2 (1991) (Luzum and Petit, 2010). The transformation between these two system accounts for three different effects:

- The motion of the celestial pole wrt. the celestial reference system (*precession* of the ecliptic and the equator, as well as *nutation*),
- the rotation of the Earth and
- the polar motion.

The transformations are typically referred to as the Celestial Intermediate Origin (CIO) approach, which has been defined in the IAU 2000/2006 resolutions (Luzum and Petit, 2010). In particular, this means that the IAU-2000 Nutation theory and the IAU-2006 Precession, the latter based on the *P03 model* (Capitaine et al., 2003; Wallace and Capitaine, 2006), are the recommended models. The conversion uncertainty is on the order of magnitude of milliarcseconds (mas), with 1 mas corresponding to a displacement of 3 cm at a distance of one Earth radius (Coppola et al., 2011).

The realisation of the ITRS, the International Terrestrial Reference Frame (ITRF), is regularly revised, with the axes definitions based on a weighted combination of a varying number of precisely known station coordinates. Since 1984, twelve versions have been published, from ITRF88 to ITRF2008 (Luzum and Petit, 2010). Using the Earth Orientation Parameters (EOP) provided and regularly updated by the IERS for the current realisation, ITRF2008, this frame is the Earth-fixed frame used in this thesis, wherever the ITRF or a body-fixed frame are referenced.

The realisation of the ICRS, the International Celestial Reference Frame (ICRF), is defined for the barycentre of the solar system. The Geocentric Celestial Reference Frame (GCRF) has the same orientation as the ICRF, but has its origin at the centre of mass of the Earth. The axes are realised from Very Long Baseline Interferometry (VLBI) observations of extragalactic radio sources (Luzum and Petit, 2010).

It should be noted that while the GCRF and ITRF and the IAU 2000/2006 conversion are the IAU-recommended method for astrodynamics applications, many of the operational systems today are still using the former IAU-76/FK5 reduction, which was the IAU standard until 1998 (Vallado et al., 2006b). In Figure 2.3, both reduction[2] methods are compared with each other. Both inertial frames coincide for the reference epoch J2000.0 with an error between the FK5 pole and the ICRS pole of ± 50 mas (Luzum and Petit, 2010).

Unless stated otherwise, the frame conversions performed for the analyses in this thesis were based on the IAU 2000/2006 resolutions.

[2]The series of translations and rotations relating the terrestrial to a celestial frame are referred to as **reduction formulas.**(Seidelmann, 2006)

Conversion between GCRF and ITRF

The transformation between the GCRF and the ITRF is accomplished via three consecutive rotations:

$$\mathbf{r}_{GCRF} = \mathbf{Q}(t) \cdot \mathbf{R}(t) \cdot \mathbf{W}(t) \cdot \mathbf{r}_{ITRF} = \mathbf{T}_{ITRF}^{GCRF} \cdot \mathbf{r}_{ITRF}, \tag{2.21}$$

where $\mathbf{Q}(t)$ is the combined bias-precession-nutation matrix, $\mathbf{R}(t)$ is accounting for the Earth's rotation and $\mathbf{W}(t)$ for the polar motion. The time parameter t used in the conversion, according to Luzum and Petit (2010), is defined as the number of Julian centuries (in Terrestrial Time (TT)) since the date 2000 January 1.5, while t_{TT} is the current Julian day in TT:

$$t = \frac{t_{TT} - 2451545.0\ TT}{36\,525} \tag{2.22}$$

Terrestrial time (TT), sometimes also referred to as Terrestrial Dynamical Time (TDT), is the "independent argument for apparent geocentric ephemerides" (Seidelmann, 2006). It uses the SI second and has a constant offset to International Atomic Time (*Temps Atomique International*) (TAI):

$$TT = TAI + 32.184^s \tag{2.23}$$

The individual steps for the transformation, as well as the intermediate frames called Celestial Intermediate Reference Frame (CIRF) and Terrestrial Intermediate Reference Frame (TIRF), are shown in Figure 2.3.

The required steps to convert from the GCRF to the ITRF are shown again in Figure 2.4, where the intersection between the two principal planes, the ecliptic and the equator, is shown for all intermediate steps for the indicated date. The precession-nutation step is separated into three single rotations by Vallado et al. (2006b), where the quantities E and d are:

$$E = \arctan \frac{Y}{X} \tag{2.24}$$

$$d = \arctan \sqrt{\frac{X^2 + Y^2}{1 - X^2 - Y^2}} \tag{2.25}$$

The reduction from the Earth-fixed frame (ITRF) to the intertial Earth-centered frame (GCRF), according to the IAU 2000/2006 resolution, includes three steps, with more details given in Annex C):

1. The **polar motion**, which is the difference between the Celestial Intermediate Pole (CIP) and the IERS Reference Pole (IRP), comprises three consecutive rotations. The first two reduce the polar coordinates x_p and y_p, which are the measured polar coordinates of the CIP in the ITRF. The third rotation, which is only required for the IAU 2000/2006 approach, is around the z-axis, with the angle s' being the so-called Terrestrial Intermediate Origin (TIO) locator, that "provides the position of the TIO

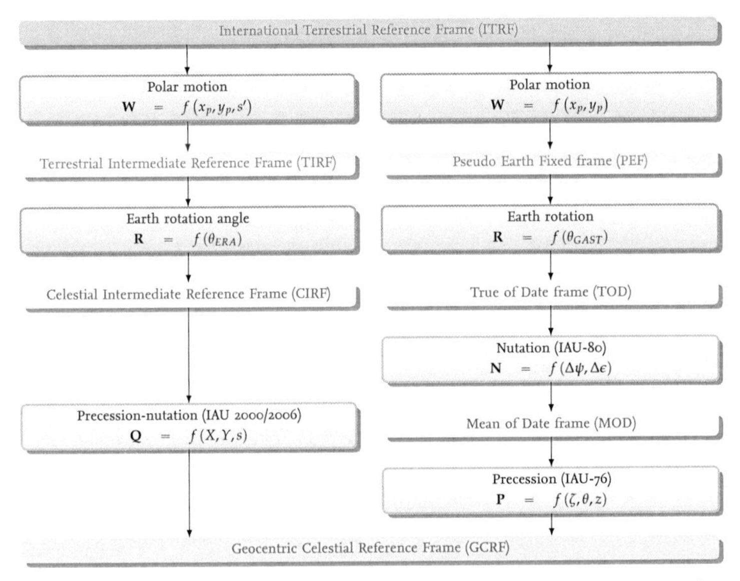

Figure 2.3.: Reduction of the terrestial coordinates. On the left, the IAU 2000/2006 method (Luzum and Petit, 2010) and on the right the former recommendation, IAU-76/FK5 (McCarthy, 1996).

on the equator of the CIP corresponding to the kinematical definition of the 'non-rotating' origin (NRO) in the ITRS when the CIP is moving with respect to the ITRS due to polar motion." (Luzum and Petit, 2010).

2. The **Earth rotation angle** (ERA) accounts for the sidereal rotation of the Earth, being the angle between CIO and TIO and defining Universal Time corrected for polar motion (UT1) by convention (Luzum and Petit, 2010). It consists of a single rotation around the CIP.

3. The **precession** and the **nutation** theories describe the motion of the CIP in the GCRF. The combined effect is described in the IAU 2000/2006 approach by the quantities X and Y, as well as s, the "CIO locator" which provides the position of the CIO on the equator of the CIP (Luzum and Petit, 2010).

From a computational point of view, the series representation (Annex C.1) of the variables X, Y and s is quite demanding, as they have to be re-computed at every force model evaluation during the numerical integration. A very promising approach, as demonstrated by Coppola et al. (2011), is to use tabulated values for those quantities resulting in a significantly reduced computational burden. This approach has also been implemented in NEPTUNE (Lange, 2014). Exemplary results for propagation times are shown in Table 2.1.

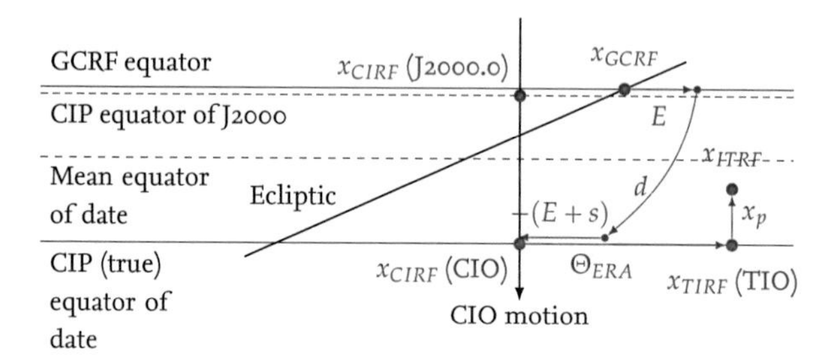

Figure 2.4.: Visualisation of the transformation from celestial to inertial frame according to the IAU 2000/2006 method (Vallado et al., 2006b).

Table 2.1.: Propagation time for different orbits with NEPTUNE, comparing full series computation of the IAU 2000/2006 CIO approach versus interpolated results. The full force model included 36×36 geopotential, atmosphere, luni-solar gravity, SRP, tides and albedo. Computations performed on an Intel® Core™ i7 (2.67 GHz).

Configuration	h_p / km	e	i / deg	Series / s	Interp. / s
LEO, full, 1 week	400.0	0.0	54.0	10.1	4.6
LEO, full, 1 month	800.0	0.0	98.6	37.4	18.0
GTO, full, 1 month	400.0	0.723	28.0	7.0	3.1
GEO, full, 1 year	35 786.0	0.0	0.0	9.0	4.1
LEO, 2-body, 1 week	400.0	0.0	54.0	1.7	0.1
LEO, 2-body, 1 month	800.0	0.0	98.6	7.7	0.3
GTO, 2-body, 1 month	400.0	0.723	28.0	3.6	0.2
GEO, 2-body, 1 year	35 786.0	0.0	0.0	4.3	0.2

For a full force model, using a 5th order Lagrange interpolation for the tabulated data results in half the time required for the full series computation. For a two-body propagation, the computation time even reduces by a factor of about 20 using tabulated and interpolated data.

2.3. Gravitational perturbations

Newton's law of universal gravitation describes the relative motion of celestial bodies under the influence of gravity. It is valid for point masses, but can also be applied for spherically symmetric bodies with uniform mass distribution. For arbitrary shaped objects, one has to consider gravitational perturbations, which can be formulated as a superposition on Newton's law (Equation 2.1).

For Earth-orbiting objects, the main characteristic of gravitational perturbations, as opposed to non-gravitational perturbations described in Section 2.4, is that one can approximate the satellites and debris objects as point masses, which results in a minimum number of parameters required to describe them.

Gravitational perturbations include the non-spherical shape of the Earth (the geopotential), solid and ocean Earth tides, the gravitational attraction of the Sun, Moon and the solar system planets or the General relativity correction.

While the geopotential is the major perturbation for most of the objects in the LEO up to the Geostationary Earth Orbit (GEO) region, especially due to the effect of Earth's oblateness, lunisolar gravitational perturbations are of particular importance for orbits at higher altitudes. Tides, on the other hand, have to be considered, as soon as the orbit determination process results in solutions better than a few tens of metres in position accuracy at epoch, while the general relativity correction is relevant in the sub-metre regime.

2.3.1. Non-spherical geopotential of the Earth

The non-spherical part of the geopotential can be formulated as an infinite sum of spherical harmonics with degree n and order m as a function of the geocentric radius r, the geocentric latitude ϕ_{gc} and the longitude λ (Kaula, 2000):

$$U\left(r,\phi_{gc},\lambda\right) = \frac{\mu}{r}\sum_{n=2}^{\infty}\sum_{m=0}^{n}\left(\frac{R_{\oplus}}{r}\right)^{n}\left(C_{n,m}\cos\left(m\lambda\right) + S_{n,m}\sin\left(m\lambda\right)\right)P_{n,m}\left(\sin\phi_{gc}\right) \tag{2.26}$$

Stokes coefficients

The spherical harmonic coefficients (or Stokes coefficients) $C_{n,m}$ and $S_{n,m}$ result from an iterative least squares fit on a set of gravity field measurements, which are obtained from satellite tracking measurements and/or surface gravity data (Förste et al., 2008). In Fig-

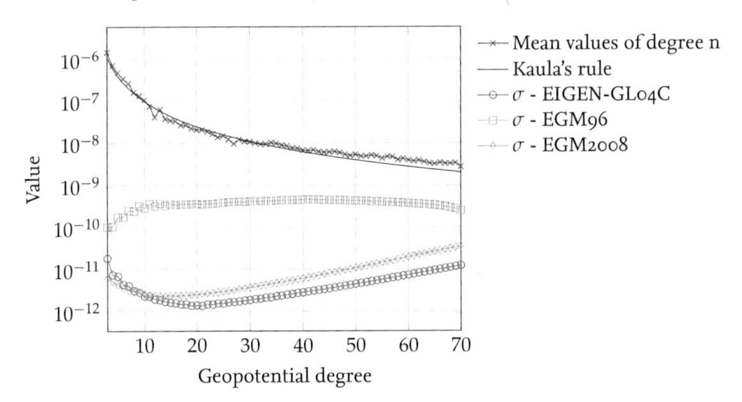

Figure 2.5.: Mean value of the normalised spherical harmonic coefficients, as well as the normalised standard deviations for different gravity models as a function of the geopotential degree.

ure 2.5, the mean values μ_n of the normalised harmonic coefficients as a function of degree n are shown, as given by Kaula (2000):

$$\mu_n = \sqrt{\frac{1}{2n+1} \sum_{m=0}^{n} \left(\overline{C}_{n,m}^2 + \overline{S}_{n,m}^2 \right)}. \tag{2.27}$$

Kaula (2000) also provided an approximation to estimate the mean value for any degree n:

$$\mu_n \approx \frac{10^{-5}}{n^2}. \tag{2.28}$$

In addition, the mean standard deviations of degree n are shown in Figure 2.5 for the three different gravity models implemented in NEPTUNE . The degree error variances are computed as (Lemoine et al., 1998):

$$\sigma_n^2 = \sum_{m=0}^{n} \left(\sigma^2 \left(\overline{C}_{n,m} \right) + \sigma^2 \left(\overline{S}_{n,m} \right) \right). \tag{2.29}$$

Recursive formulation for computing the Legendre functions

In the formulation of the geopotential via Equation 2.26, the associated Legendre functions, $P_{n,m}$, need to be evaluated. They can be computed directly according to their definition given in Equation 2.30, but this is a laborious task, even for computers.

$$P_{n,m}(x) = (-1)^m \left(1 - x^2\right)^{m/2} \frac{d^m}{dx^m} P_n(x). \tag{2.30}$$

A much easier way to obtain the results for given degree and order, is to use recursions. According to Montenbruck and Gill (2000), using the relation $x = \sin \phi_{gc}$:

$$P_{m,m}(x) = (2m-1)\sqrt{1-x^2} P_{m-1,m-1} \tag{2.31}$$

$$P_{m+1,m}(x) = (2m+1)\, x P_{m,m}(x) \tag{2.32}$$

$$P_{n,m}(x) = \frac{1}{n-m}\left((2n-1)\, x P_{n-1,m}(x) - (n+m-1)\, P_{n-2,m}(x)\right), \ \forall n > m+1, \tag{2.33}$$

with the required initial values to start the iteration:

$$P_{0,0}(x) = 1 \tag{2.34}$$

$$P_{1,0}(x) = x = \sin \phi_{gc} \tag{2.35}$$

$$P_{1,1}(x) = \sqrt{1-x^2} = \cos \phi_{gc}. \tag{2.36}$$

An important point is that a recurrence formulation may become instable after many consecutive evaluations. This happens, for example, if differences of two nearly equal numbers are processed, or if there are small divisors (Vallado and McClain, 2013).

In order to have a stable recurrence for the Legendre polynomials, a scheme as shown in Figure 2.6 has to be followed with Equation 2.31 through Equation 2.33 (Montenbruck

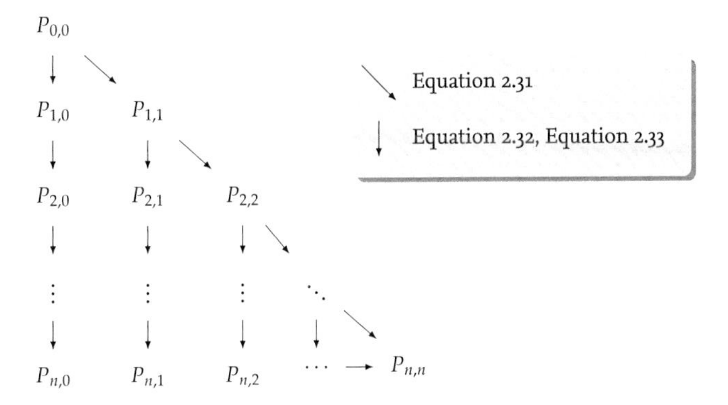

Figure 2.6.: Using a recurrence formulation for the Legendre functions.

and Gill, 2000; Brumberg, 1995). An interesting analysis on the stability of different recurrence formulations for the Legendre functions was done by Lundberg (1985). The analysis showed that, especially for high degree computations, the recurrence formulation in Equation 2.33 was superior to other formulations in terms of stability.

Just to show an example, Figure 2.7 gives the difference between the approach as described above, i.e. using Equation 2.31 through Equation 2.33 in combination with the scheme shown in Figure 2.6, and a different recurrence formulation (e.g. as given by Long et al. (1989)), without a specific scheme:

$$P_{n,0}(x) = \frac{1}{n}\left((2n-1)\,xP_{n-1,0}(x) - (n-1)\,P_{n-2,0}(x)\right),\; n \geqslant 2 \tag{2.37}$$

$$P_{n,m}(x) = P_{n-2,m}(x) + (2n-1)\sqrt{1-x^2}P_{n-1,m-1}(x),\; m \neq 0, m < n \tag{2.38}$$

$$P_{n,n}(x) = (2n-1)\sqrt{1-x^2}P_{n-1,n-1}(x),\; n \neq 0 \tag{2.39}$$

It can be seen in Figure 2.7 that selecting a typical geopotential degree between 30 and 40 may already introduce significant errors, which indicate that from 15 available digits only 7 or 8 actually contain useful information. The latter is especially true for those terms, where $m \approx \frac{n}{2}$, while there is no difference between the two methods for zonal ($m = 0$) or sectorial ($n = m$) terms. It can also be noted that for even higher degrees the error tends to grow exponentially.

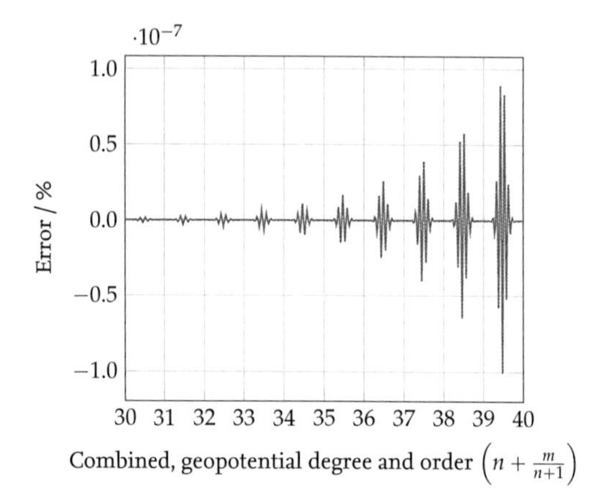

Figure 2.7.: Relative error between two different recurrence computation methods for the geopotential degree being between 30 and 40. The first method used Equation 2.31 through Equation 2.33 in combination with the scheme shown in Figure 2.6, which is considered to be stable according to Montenbruck and Gill (2000); Brumberg (1995); Lundberg (1985). The second method uses the recurrence formulation given by Equation 2.37 through Equation 2.39. Both, the geopotential degree and order were combined as the independent variable x according to $x = n + \frac{m}{n+1}$.

Accelerations in body-fixed frame

The accelerations of the non-spherical geopotential, which are required for the evaluation of Equation 2.1, are obtained by differentiating the potential with respect to the radius vector, the latter given in a body-fixed frame:

$$\mathbf{a}_{bf} = \frac{\partial U}{\partial \mathbf{r}_{bf}} = \frac{\partial U}{\partial r}\left(\frac{\partial r}{\partial \mathbf{r}_{bf}}\right)^T + \frac{\partial U}{\partial \phi_{gc}}\left(\frac{\partial \phi_{gc}}{\partial \mathbf{r}_{bf}}\right)^T + \frac{\partial U}{\partial \lambda}\left(\frac{\partial \lambda}{\partial \mathbf{r}_{bf}}\right)^T \tag{2.40}$$

The result obtained from Equation 2.40 is the inertial acceleration of the satellite expressed in an Earth-fixed frame, so that only the rotations to an inertial reference frame are required (Long et al., 1989):

$$\mathbf{a}_{GCRF} = \mathbf{T}_{ITRF}^{GCRF}\mathbf{a}_{ITRF} \tag{2.41}$$

The partial derivatives for the evaluation of Equation 2.40 are given by Vallado and McClain (2013):

$$\frac{\partial U}{\partial r} = -\frac{\mu}{r^2}\sum_{n=2}^{\infty}\sum_{m=0}^{n}\left(\frac{R_\oplus}{r}\right)^n (n+1)\left(C_{n,m}\cos(m\lambda)+S_{n,m}\sin(m\lambda)\right)P_{n,m}\left(\sin\phi_{gc}\right)$$

$$\frac{\partial U}{\partial \phi_{gc}} = \frac{\mu}{r}\sum_{n=2}^{\infty}\sum_{m=0}^{n}\left(\frac{R_\oplus}{r}\right)^n \left(C_{n,m}\cos(m\lambda)+S_{n,m}\sin(m\lambda)\right)\times$$
$$\times\left(P_{n,m+1}\left(\sin\phi_{gc}\right)-m\tan\phi_{gc}P_{n,m}\left(\sin\phi_{gc}\right)\right)$$

$$\frac{\partial U}{\partial \lambda} = \frac{\mu}{r}\sum_{n=2}^{\infty}\sum_{m=0}^{n}\left(\frac{R_\oplus}{r}\right)^n mP_{n,m}\left(\sin\phi_{gc}\right)\left(S_{n,m}\cos(m\lambda)-C_{n,m}\sin(m\lambda)\right) \tag{2.42}$$

Finally, the accelerations in the body-fixed frame can be computed from:

$$a_{x,bf} = \left(\frac{1}{r}\frac{\partial U}{\partial r}-\frac{r_z}{r^2\sqrt{r_x^2+r_y^2}}\frac{\partial U}{\partial \phi_{gc}}\right)r_x-\left(\frac{1}{r_x^2+r_y^2}\frac{\partial U}{\partial \lambda}\right)r_y-\frac{\mu r_x}{r^3} \tag{2.43}$$

$$a_{y,bf} = \left(\frac{1}{r}\frac{\partial U}{\partial r}-\frac{r_z}{r^2\sqrt{r_x^2+r_y^2}}\frac{\partial U}{\partial \phi_{gc}}\right)r_y+\left(\frac{1}{r_x^2+r_y^2}\frac{\partial U}{\partial \lambda}\right)r_x-\frac{\mu r_y}{r^3} \tag{2.44}$$

$$a_{z,bf} = \frac{1}{r}\frac{\partial U}{\partial r}r_z+\frac{\sqrt{r_x^2+r_y^2}}{r^2}\frac{\partial U}{\partial \phi}-\frac{\mu r_z}{r^3} \tag{2.45}$$

2.3.2. Third body gravitational perturbations

The gravitational perturbations due to solar system bodies are governed mainly by the Sun and the Moon for Earth-orbiting objects. The acceleration of a so-called third body can be directly expressed with Newton's gravitational law:

$$\mathbf{a}_{3b} = GM_{3b}\left(\frac{\mathbf{r}_{3b}-\mathbf{r}_S}{\left|\mathbf{r}_{3b}-\mathbf{r}_S\right|^3}-\frac{\mathbf{r}_{3b}}{\left|\mathbf{r}_{3b}\right|^3}\right), \tag{2.46}$$

with $\mathbf{r}_{3b}$ and $\mathbf{r}_S$ being the geocentric radius vector of the third body and the satellite, respectively.

For most applications, it is sufficient to consider only lunisolar contributions, as the accelerations due to the other celestial bodies are smaller by several orders of magnitudes. For example, the acceleration ratio Sun to Jupiter is:

$$\frac{a_\odot}{a_{♃}} = \frac{M_\odot}{M_{♃}}\frac{r_{♃}^3}{r_\odot^3} \approx 10^5. \tag{2.47}$$

For a circular LEO at 400 km altitude and 54° inclination, the additional error introduced by Jupiter in the propagation would be on the order of 10 m after one day.

The ephemerides of solar system bodies are publicly available via the Development Ephemerides (DE) series provided by the Jet Propulsion Laboratory (JPL)[3]. A convenient way is to use the Spacecraft Planet Instrument C-matrix Events (SPICE) toolkit provided by NASA's Navigation and Ancillary Information Facility (NAIF). The ephemerides are recovered from a SPICE Kernel (SPK) file in the GCRF. The specific gravitational constants for the solar system bodies are also provided in order to be consistent with the polynomial series.

2.3.3. Tides

The spherical harmonics coefficients (Section 2.3.1) are, in general, provided for a tide free geopotential. This means that the observed instantaneous crust is reduced by removing total tidal deformations, resulting in a tide free crust, which is also the reference for the ITRF (Luzum and Petit, 2010). Solid Earth tides result in accelerations comparable to the geopotential contributions of the spherical harmonics at degree $n \approx 15$ and can be even more important than solar radiation pressure in the LEO regime (Montenbruck and Gill, 2000).

Solid Earth tides

The changes in the geopotential as induced by solid Earth tides, are modelled by introducing variations to the spherical harmonics coefficients $C_{n,m}$ and $S_{n,m}$ (Luzum and Petit, 2010). Using the frequency-independent nominal Love numbers k_{nm}, the solid Earth tide contributions can be computed from (Luzum and Petit, 2010):

$$\begin{pmatrix} \Delta\overline{C}_{nm} \\ \Delta\overline{S}_{nm} \end{pmatrix} = \frac{k_{nm}}{2n+1}\left(\frac{GM_{3b}}{GM_{\oplus}}\right)\left(\frac{R_{\oplus}}{r_{3b}}\right)^{n+1}\overline{P}_{nm}\left(\sin\phi_{3b}\right)\begin{pmatrix} \cos\left(m\lambda_{3b}\right) \\ \sin\left(m\lambda_{3b}\right) \end{pmatrix}. \tag{2.48}$$

The above equation has to be evaluated twice, to obtain the solar and lunar contributions, respectively. The normalised Legendre functions are computed for the geocentric latitude and longitude of both the Sun $(\phi_{\odot}, \lambda_{\odot})$ and the Moon $(\phi_{☾}, \lambda_{☾})$.

Note that Equation 2.48 provides the variations of the normalised coefficients, while the geopotential might be implemented for computations based on unnormalised values. The following relationship is used for the (de-)normalisation:

$$\Pi_{nm} = \sqrt{\frac{(n+m)!}{(2-\delta_{0m})(2n+1)(n-m)!}} \tag{2.49}$$

where δ_{0m} is the Kronecker delta, with $\delta_{0m} = 1$ if $m = 0$ and zero otherwise. The normalisation is then achieved by:

$$\begin{pmatrix} \overline{C}_{nm} \\ \overline{S}_{nm} \end{pmatrix} = \Pi_{nm}\begin{pmatrix} C_{nm} \\ S_{nm} \end{pmatrix}, \quad \text{and} \quad \overline{P}_{nm} = \frac{P_{n,m}(x)}{\Pi_{nm}}. \tag{2.50}$$

[3]http://ssd.jpl.nasa.gov/?horizons, accessed on September 29, 2015.

Substituting Equation 2.50 into Equation 2.48 one obtains the corrections to the unnormalised coefficients:

$$\begin{aligned}\begin{pmatrix}\Delta C_{n,m}\\ \Delta S_{n,m}\end{pmatrix} &= \begin{pmatrix}\Delta\overline{C}_{nm}\\ \Delta\overline{S}_{nm}\end{pmatrix}\frac{1}{\Pi_{nm}}\\ &= \frac{1}{\Pi_{nm}^2}\frac{k_{nm}}{2n+1}\left(\frac{GM_{3b}}{GM_\oplus}\right)\left(\frac{R_\oplus}{r_{3b}}\right)^{n+1}P_{nm}\left(\sin\phi_{3b}\right)\begin{pmatrix}\cos\left(m\lambda_{3b}\right)\\ \sin\left(m\lambda_{3b}\right)\end{pmatrix}\\ &= k_{nm}\delta_{0m}\frac{(n-m)!}{(n+m)!}\left(\frac{GM_{3b}}{GM_\oplus}\right)\left(\frac{R_\oplus}{r_{3b}}\right)^{n+1}P_{nm}\left(\sin\phi_{3b}\right)\begin{pmatrix}\cos\left(m\lambda_{3b}\right)\\ \sin\left(m\lambda_{3b}\right)\end{pmatrix}\end{aligned} \tag{2.51}$$

Tides of second degree ($n = 2$) introduce changes in the geopotential coefficients of fourth degree. Luzum and Petit (2010) gives the corrections for the normalised values. The result for unnormalised coefficients is obtained from Equation 2.48 using the transformation from Equation 2.50:

$$\begin{aligned}\begin{pmatrix}\Delta C_{4m}\\ \Delta S_{4m}\end{pmatrix} &= \frac{1}{\Pi_{4m}\Pi_{2m}}\frac{k_{nm}^{(+)}}{5}\left(\frac{GM_{3b}}{GM_\oplus}\right)\left(\frac{R_\oplus}{r_{3b}}\right)^{3}P_{2m}\left(\sin\phi_{3b}\right)\begin{pmatrix}\cos\left(m\lambda_{3b}\right)\\ \sin\left(m\lambda_{3b}\right)\end{pmatrix}\\ &= k_{2m}^{(+)}\delta_{0m}\frac{(2-m)!}{(2+m)!}\sqrt{\frac{9\left(4-m\right)\left(3-m\right)}{5\left(4+m\right)\left(3+m\right)}}\left(\frac{GM_{3b}}{GM_\oplus}\right)\left(\frac{R_\oplus}{r_{3b}}\right)^{3}P_{2m}\left(\sin\phi_{3b}\right)\cdot\\ &\cdot\begin{pmatrix}\cos\left(m\lambda_{3b}\right)\\ \sin\left(m\lambda_{3b}\right)\end{pmatrix},\end{aligned} \tag{2.52}$$

with the above equation being evaluated for $m = 0, 1, 2$.

Solid Earth pole tide

Another correction is applied for the *pole tide*, which is due to the centrifugal effect of polar motion (Luzum and Petit, 2010). Again, the unnormalised corrections are:

$$\Delta C_{21} = -1.721\cdot 10^{-9}\left(m_1 - 0.0115 m_2\right) \tag{2.53}$$

$$\Delta S_{21} = -1.721\cdot 10^{-9}\left(m_2 + 0.0115 m_1\right) \tag{2.54}$$

The quantities m_1 and m_2 are defined as the deviations of the polar motion variables x_p and y_p from their running averages. Figure 2.8 shows the position error introduced during propagation when omitting solid Earth tides, which is on the order of several metres for a few days of propagation.

Ocean tides

The reaction of Earth's water mass to the gravitational forces of the Sun and the Moon is referred to as ocean tides, which provide a contribution of about 10 % to 15 % compared to solid Earth tides, as suggested by Casotto (1989). Due to the very complex nature of the water motion, ocean tide models are typically represented by models with a broad range of frequencies and amplitudes in the spectrum. Luzum and Petit (2010) discuss the different

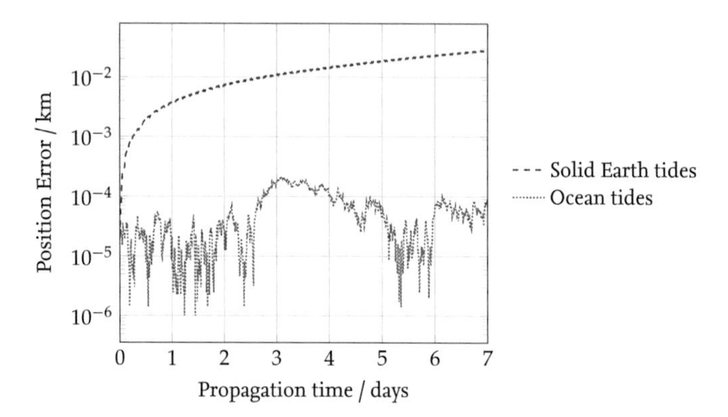

Figure 2.8.: Solid Earth and ocean tides example for an 800 km orbit ($i = 98.7°$, $e = 0.001$). The position error is computed for a trajectory comparison against a two-body propagation.

models recommended by the IERS. Luzum and Petit (2010) also give an example for the influence of ocean tides: For a satellite in an 800 km orbit ($i = 98.7°$, $e = 0.001$) and one day of propagation, the error is on the order of several cm and reaches up to 20 cm. This is also shown in Figure 2.8.

2.4. Non-gravitational perturbations

The point mass assumption is inadequate in the modeling of non-gravitational perturbations. Additional parameters are required to account for several forces due to the interaction of the orbiting object's surfaces with the environment. If thermal and optical properties of a spacecraft are known, a detailed surface model allows to account for all the effects in a precise 6-DOF propagation. For space debris and spacecraft with unknown properties, those parameters are typically solved for in the orbit determination process. Examples for such parameters are the drag coefficient c_D or the SRP coefficient c_R.

2.4.1. Atmospheric drag

The acceleration due to the object's movement in the Earth's atmosphere can be described by the drag equation:

$$\mathbf{a}_D = -\frac{\rho}{2}\frac{c_D A_c}{m}\mathbf{v}_{rel}\left|\mathbf{v}_{rel}\right|, \tag{2.55}$$

with the total density ρ, the velocity $\mathbf{v}_{rel}$relative to the Earth's atmosphere, the object's mass m, the cross-section A_c being the orthographic projection of the object's surface in the direction of $\mathbf{v}_{rel}$ and c_D being the drag coefficient.

The drag coefficient generally assumes values between $c_D = 2.0 \ldots 2.2$ under free molecular flow conditions and the above definition of the cross-section.

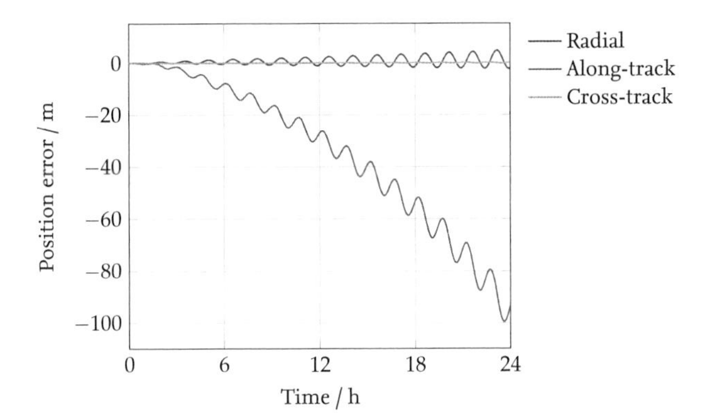

Figure 2.9.: Example for the position difference in the Orbit Centered Reference Frame (OCRF) between a force model comprised of drag (NRLMSISE-00) and wind (HWM07) compared with a drag only model. The orbit is circular at 300 km altitude with an inclination of 60°.

Typically, for debris fragments, all of the quantities, m, A_c and c_D are unknown and cannot be determined independently from observations. It is thus convenient to combine them in the so-called *ballistic coefficient*[4]:

$$B = \frac{c_D A_c}{m}. \tag{2.56}$$

The density is obtained from thermosphere models, like the widely used NRLMSISE-00 (Naval Research Laboratory Mass Spectrometer and Incoherent Scatter (Extended), Earth atmosphere model) applicable to altitude regimes down to Earth's surface (Picone et al., 2002). The empirical model accounts for long-term effects, like annual and semiannual variations, and for short-term variations on diurnal, semidiurnal and terdiurnal scale. The NRLMSISE-00 model is very convenient for propagation purposes, as it requires only few input parameters: besides the position in the ITRF, the solar acitivity proxy $F_{10.7}$ (the solar radiation at a wavelength of 10.7 cm measured on ground) and the geomagnetic activity planetary index, A_p, are required.

The relative velocity $\mathbf{v}_{rel}$ is the difference between the velocities of the object and the atmosphere. The latter, co-rotating with the Earth, is superimposed by winds, which can be modelled up to altitudes of 500 km using the Horizontal Wind Model HWM07 (Drob et al., 2008). The contributions due to wind can be on the order of 100 m/s at high latitudes (Drob et al., 2008). In Figure 2.9, an example is given for a 300 km orbit with 60° inclination, showing the influence of horizontal wind. It can be seen that the position error in the along-track component is on the order of 100 m after 24 h of propagation. The error

[4]The ballistic coefficient is often also defined in a reciprocal way: $B = \frac{m}{c_D A_c}$. It is therefore important to be aware of how it is defined when working with this quantity.

in the along-track component reduces to about 1 m for the same example, if the altitude is increased to 400 km, corresponding to the orbit of the ISS.

2.4.2. Solar radiation pressure

Solar photons absorbed by or reflected from the satellite's surfaces provide an impulse change and thus a perturbative acceleration. These forces due to SRP are of comparable magnitude to atmospheric drag at altitudes of about 600 km and are the main non-gravitational perturbation for higher orbits.

It is essential to have a model for both, the incoming solar flux at the satellite's position, as well as the surfaces and their optical properties.

The solar flux is typically given at a mean solar distance of 1 ua for the complete solar spectrum, often also referred to as the *solar constant*. The solar constant has been revised in 2011 and is given by Kopp and Lean (2011):

$$\Phi_{\odot} = (1360.8 \pm 0.5) \, \frac{\mathrm{W}}{\mathrm{m}^2} \tag{2.57}$$

The solar radiation pressure is obtained by dividing the solar constant by the speed of light:

$$P_{\odot} = \frac{\Phi_{\odot}}{c} = 4.5391 \times 10^{-6} \, \frac{\mathrm{N}}{\mathrm{m}^2}. \tag{2.58}$$

For an absorbing sphere, the resulting force is now simply computed by multiplying the solar radiation pressure with the illuminated cross-section. The general case is for a set of individually oriented surfaces, which together define the satellite macro model. The incoming photons interact with the surfaces and are either reflected (specularly or diffusely) or absorbed. The geometry, defining the unit vector $\mathbf{e}_{\odot}$ pointing to the Sun, the surface normal vector $\mathbf{n}_0$ and the solar incidence angle ϕ_{inc} between these two vectors, is shown in Figure 2.10.

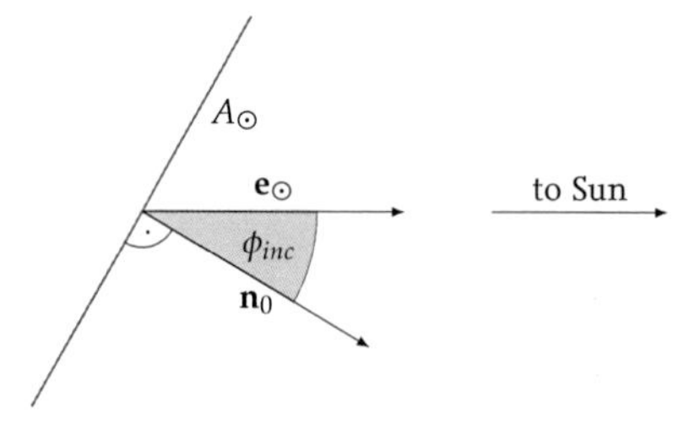

Figure 2.10.: Definition of the incidence angle for oriented surfaces.

The force due to absorption by a surface of area A is in the Sun's direction and thus equal to:

$$\mathbf{F}_{abs} = -P_{\odot} c_A A \cos \phi_{inc} \mathbf{e}_{\odot} \tag{2.59}$$

For a specular reflection, no impulse is transferred parallel to the surface. The resulting force is only in surface normal direction. Due to the specular reflection, the impulse in normal direction is twice the initial impulse:

$$\mathbf{F}_{r,s} = -2P_{\odot}c_{R,s}A\cos^2\phi_{inc}\mathbf{n}_0 \tag{2.60}$$

The cosine function is squared to first account for the orientation of the surface with respect to the Sun and second for the momentum transfer in normal direction. The diffuse reflectivity is modelled by assuming a Lambert distribution (Klinkrad and Fritsche, 1998; Vallado and McClain, 2013):

$$\mathbf{F}_{r,d} = -P_{\odot}c_{R,d}A\cos\phi_{inc}\left(\frac{2}{3}\mathbf{n}_0 + \mathbf{e}_{\odot}\right) \tag{2.61}$$

The absorption coefficent c_A, specular reflectivity coefficient $c_{R,s}$ and diffuse reflectivity coefficient are interrelated:

$$c_A + c_{R,s} + c_{R,d} = 1 \tag{2.62}$$

Combining the above equations and summing over all illuminated surfaces, one finally obtains the acceleration due to solar radiation pressure in the inertial frame:

$$\mathbf{a}_{SRP} = -\frac{P_{\odot}}{m}\left(\frac{1;AU}{\mathbf{r}_{S\odot}}\right)^2\sum_{i=1}^{n_{srf}} A_i\cos\phi_{inc,i}\left(2\left(\frac{c_{R,d,i}}{3} + c_{R,s,i}\cos\phi_{inc,i}\right)\mathbf{n}_{0,i} + (1 - c_{R,s,i})\,\mathbf{e}_{\odot,i}\right) \tag{2.63}$$

Shadow model

For many orbits it is important to consider eclipse conditions, which occur when the Sun gets occulted by the Earth. In some cases, also effects of the lunar eclipses can be considered for high-fidelity modeling (Escobal and Robertson, 1967; Srivastava et al., 2015). While analytical and semi-analytical theories often involve a cylindrical umbra-only (fully occulted solar disk) shadow region to allow for an analytical computation of the boundary crossings (e.g. Aksnes (1976)), in numerical integration it is convenient to use a conical model with umbra and penumbra (partial occultation) regions, as shown in Figure 2.11. The acceleration due to solar radiation pressure, as given by Equation 2.63, is multiplied with the shadow function ν, which is either $\nu = 1$ in full sunlight, $\nu = 0$ in the umbra, and assumes values between 0 and 1 in the penumbra region (Montenbruck and Gill, 2000).

For the geometry shown in Figure 2.11b, Montenbruck and Gill (2000) give a simple method to compute the shadow function value.

$$\nu = 1 - \frac{A}{\pi a^2}, \tag{2.64}$$

with A being the occulted segment of the apparent solar disk and a the apparent radius of the Sun.

The trigonometric function to compute A results in what is referred to and shown as the *true* shadow function in Figure 2.12. The shadow function was evaluated for an

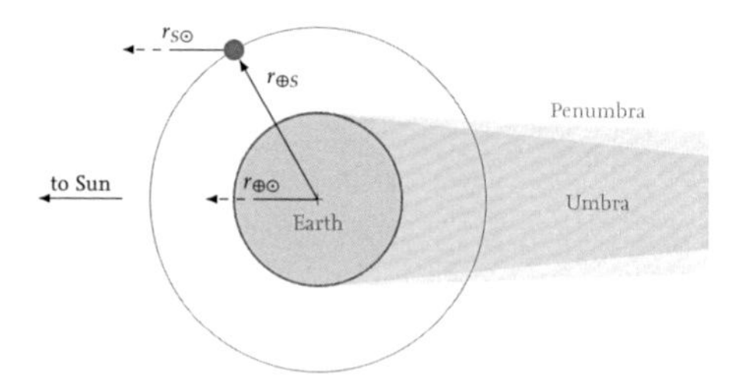

(a) Umbra and Penumbra

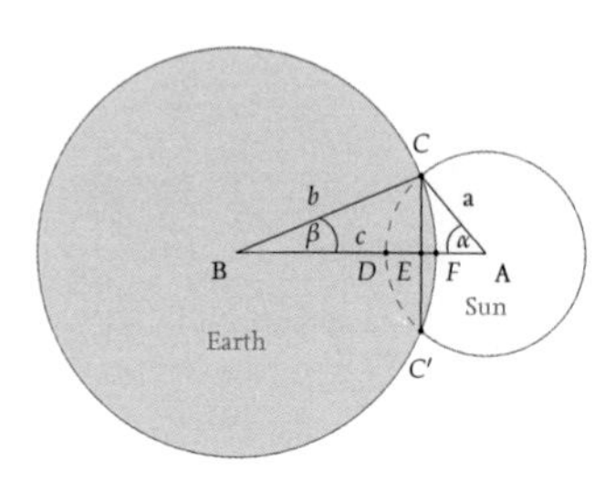

(b) Sun occultation by the Earth

Figure 2.11.: Conical shadow model with umbra and penumbra (left) and the occultation of the Sun by the Earth as seen from the satellite (right).

occultation happening in the direction of the connecting line of the two bodies. In Figure 2.12, one can also see that it is convenient to assume a linear transition from full sunlight to eclipse condition across the penumbra region. The latter is the main assumption for the SBTC method (Lundberg et al., 1991), which has also been implemented in NEPTUNE (Horstmann et al., 2015) and computes state vector corrections via a simple integral of the triangular area below the linear shadow function. In fact, the difference of the integrals for the *true* shadow function and the linear approximation are very small: for Earth orbits they are always below 0.5 %, as shown in Table 2.2. As the shadow function is a linear scaling factor (between 0 and 1) for the acceleration, the errors for the latter are also those shown in Table 2.2.

While the shadow models described so far all assume a spherical Earth without an atmosphere, effects due to Earth's flattening and the refraction and absorption of Sun light by the Earth's atmosphere, causing a different transition as seen from the satellite from full sunlight to eclipse, have to be considered for precise applications. Both effects are

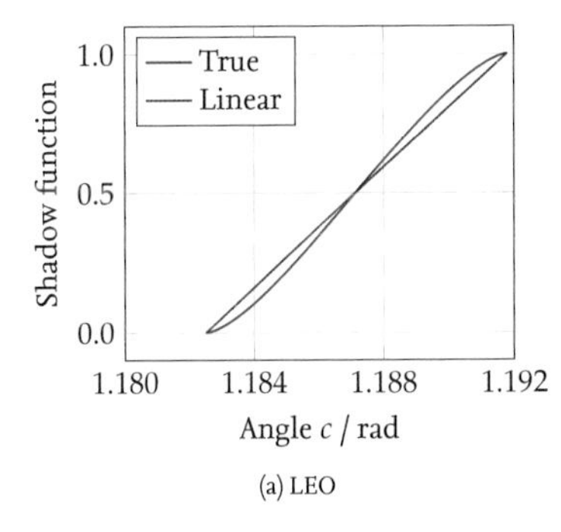

(a) LEO

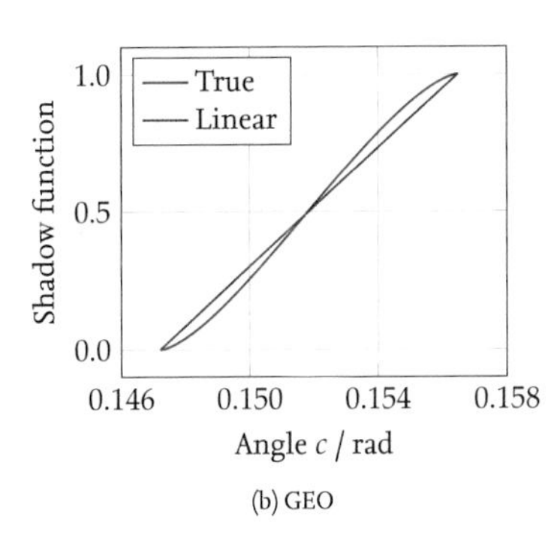

(b) GEO

Figure 2.12.: Real shadow function for circular cross-sections of occulting bodies compared with a linear penumbra model in LEO and GEO region.

Table 2.2.: Integral error for the comparison of the linear shadow function with the model for circular cross-sections of the occulting bodies. The occultation happens in the direction along the connecting line of the two bodies centers of mass.

Orbit radius / km	Error / %
6878	-0.05
8378	-0.07
26 000	-0.23
42 164	-0.38
384 000	-3.34

comparable in a certain sense: the refraction is due to an atmospheric layer of a few tens km, which is in the same order of magnitude as the deviation of Earth's figure from a sphere (Vokrouhlicky et al., 1996).

A flattened Earth causes a time shift in the penumbra entry and exit and leads to a shorter eclipse in total. However, as was analyzed by Vokrouhlicky et al. (1996), this leads only to short-periodic variations in the along-track direction, where the authors showed that for the sample case of LAGEOS (Laser Geodynamics Satellites) the displacements are on the order of 1 cm (Vokrouhlicky et al., 1996). The assumption of a spherical Earth for modeling of the SRP is thus justified in the space surveillance and tracking context, while it has to be considered for applications requiring very precise orbit solutions.

Due to the atmospheric refraction, the satellite will see the full solar disk for a longer time. The latter is distorted and part of the solar radiation is absorbed during the passage through Earth's atmosphere, which results in a complex transition process. These effects were also rigorously analyzed by Vokrouhlicky et al. (1993) and for LAGEOS they estimated a change in semi-major axis of about 1.7 cm over a shadow-crossing interval of about 100 days - compared to a spherical Earth model without atmosphere.

2.4.3. Earth radiation pressure

Another important source of radiation affecting the orbit of a satellite, is coming from the Earth. It can be subdivided into two distinct wavebands: a shortwave contribution in the visible regime from reflected sunlight (also referred to as *Albedo*), and emitted radiation from the Earth in the Infrared (IR) regime. While the former is of a very complex nature, with the satellite's exposure to lit and unlit surfaces of the Earth, the latter is nearly invariant with respect to the illumination conditions on Earth (Knocke, 1989).

At very low altitudes between 200 km to 300 km the magnitude of Earth Radiation Pressure (ERP) accelerations is about 35 % of the SRP accelerations (Knocke, 1989). For increasing altitude, this value decreases, so that Knocke (1989) gives a range between 10 % to 25 % of the direct SRP magnitude for most Earth satellites.

A model to obtain the perturbative effect on the satellite's orbit due to ERP consists of the following steps: The incident solar radiation on a surface element of the Earth is subject to either specular or diffusive reflection (or a combination thereof) as a function of the surface properties and the incidence angle. While snow-covered areas, the seas and clouds can be well approximated as specularly reflecting elements, especially at high solar zenith angles, land masses are typically of a more diffusive nature (Barlier et al., 1986). The reflected radiation is then interacting with the satellite's surfaces, in a similar way as shown for the SRP in Equation 2.63.

Different approaches exist to model the reflective interaction of incident sunlight with the Earth, ranging from assuming a perfectly specular reflection of the whole visible Earth (Vokrouhlicky et al., 1994), a zonal albedo model for the visible and IR regime (Knocke, 1989), up to a high-fidelity modeling with spherical harmonics in latitude and longitude (Sehnal, 1979).

Vokrouhlicky et al. (1994) and Knocke (1989) have shown that even simplified models can appropriately describe the observed semi-major axis variations in the orbit of LAGEOS.

Therefore, the model given by Knocke (1989) shall be briefly outlined here as a suitable method in the space surveillance context.

The visible cap of the Earth is subdivided into plates with the local normal pointing in the radial direction at the geographical coordinates of the plate's center. A central spherical cap is modelled directly below the satellite, while the other plates are distributed on two ring segments around that cap, as shown in Figure 2.13. While the ring segments may be defined in an arbitrary way, Knocke (1989) introduces the concept of *equal projected attenuated areas*. This area is defined as:

$$A_{prj,i} = \frac{dA_i \cos \alpha_{ik}}{\pi \left| \mathbf{s}_{ik} \right|}, \tag{2.65}$$

where A_i is the surface area of the i-th Earth plate, α_{ik} is the view angle from the plate i to the satellite surface element k, and $\mathbf{s}_{ik}$ is the distance.

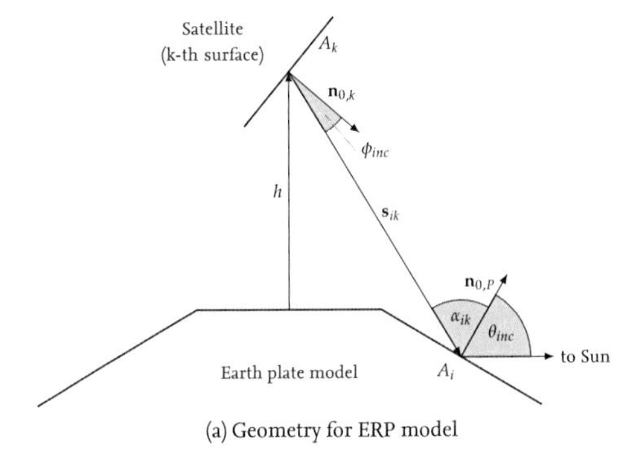

(a) Geometry for ERP model

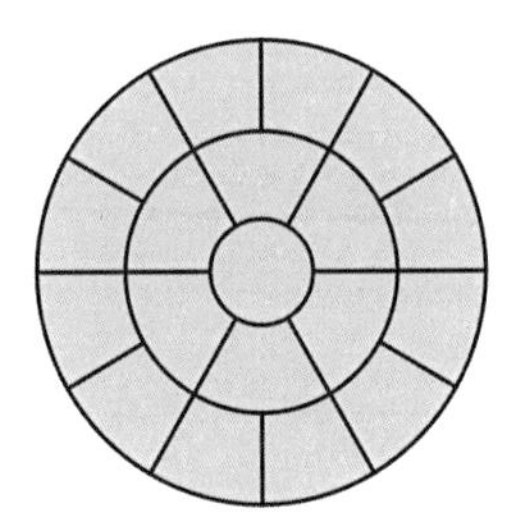

(b) Cap and ring segments seen by satellite

Figure 2.13.: Modelling Earth radiation pressure according to Knocke (1989). The visible cap is further subdivided into a circular surface directly below the satellite and concentric ring elements, where all Earth surface elements have the same projected *attenuated surface area*.

Using the projected attenuated areas, which are equal for the individual segments as well as for the central cap, the computation of the resulting acceleration simplifies significantly, resulting in (Knocke, 1989):

$$\mathbf{a}_{ERP} = K(a,e) \sum_{i=1}^{n_{seg}} \sum_{k=1}^{n_{srf}} A_{prj,i} A_k \cos\phi_{inc} \left(2 \left(\frac{c_{R,d,k}}{3} + c_{R,s,k} \cos\phi_{inc,i} \right) \mathbf{n}_{0,k} + (1 - c_{R,s,k})\, \mathbf{s}_{i,k} \right). \tag{2.66}$$

Note that the formulation in Equation 2.66 makes use of the same satellite model already introduced in the previous section. The contributions of the individual ring segments and the central caps are evaluated for each surface element of the satellite. The incoming radiation from the surface element of the Earth is described by the function:

$$K(a,e) = \frac{P_{\odot}}{m}\left(\kappa a \cos\theta_{inc} + \frac{e}{4}\right), \tag{2.67}$$

where a is the albedo and e the emissivity of the surface element. The multiplier κ is either one, if the surface element is illuminated by the Sun and zero otherwise. Both, the albedo and emissivity are given by Knocke (1989) as functions of the geocentric altitude ϕ_{gc} and time:

$$a = a_0 + a_1(t)\,P_1(\sin\phi_{gc}) + a_2 P_2(\sin\phi_{gc}) \tag{2.68}$$

$$e = e_0 + e_1(t)\,P_1(\sin\phi_{gc}) + e_2 P_2(\sin\phi_{gc}), \tag{2.69}$$

where

$$a_1(t) = c_0 + c_1\cos\left(2\pi\frac{t-t_0}{T_{\oplus}}\right) + c_2\sin\left(2\pi\frac{t-t_0}{T_{\oplus}}\right) \tag{2.70}$$

$$e_1(t) = k_0 + k_1\cos\left(2\pi\frac{t-t_0}{T_{\oplus}}\right) + k_2\sin\left(2\pi\frac{t-t_0}{T_{\oplus}}\right). \tag{2.71}$$

In the above equations, P_n are the nth degree Legendre polynomials, $T_{\oplus}$ is the orbital period of the Earth ($T_{\oplus} = 365.25\,\mathrm{d}$), t_0 is the epoch of the periodic terms and a_0, a_2, e_0, e_2, c_i and k_i are the parameters for this model. It accounts for seasonal effects and purely zonal variations.

An example, showing the influence of radiative accelerations on a real orbit, is shown in Figure 2.14 for POE obtained for the Jason-1 [5] satellite.

For the precise orbit determination of the altimetry mission a goal of 1 cm radial Root Mean Square (RMS) (and total RMS in the cm-regime, Luthcke et al. (2003)) was achieved by combining GPS, SLR and DORIS measurements. The selected orbit at 1336 km reduces the effects of atmospheric drag and the geopotential. The inclination of 66° and the fact that the orbit is not sun-synchronous, minimizes the effects of repeating tidal frequencies.

The POEs of Jason-1 were used in the validation process of Neptune. In order to perform the orbit comparison appropriately, one has to make sure that the force models of the Precise Orbit Determination (POD) and Neptune are aligned and the optical and thermal properties of the satellite are reasonably modelled. While the former may be achieved, the latter is the complicated part - thus, it was decided to perform the POE-based validation in a more qualitative way based on a cannon-ball model with estimated coefficients. The dimensions of the box-shaped satellite with two solar array wings are given in Figure 2.14a. An average cross-section of 10 m², a drag coefficient of $c_D = 2.2$ and an SRP coefficient of $c_R = 1.3$ were assumed, and the effects of solar and Earth radiation pressure analysed, as shown in Figure 2.14b through Figure 2.14d (the reference epoch in all plots is March 24, 2002). It can be seen that if both, SRP and ERP are switched off

[5] obtained from the LEGOS-CLS via the International DORIS Service (IDS) website: `http://ids-doris.org/welcome.html`, accessed on October 27, 2015.

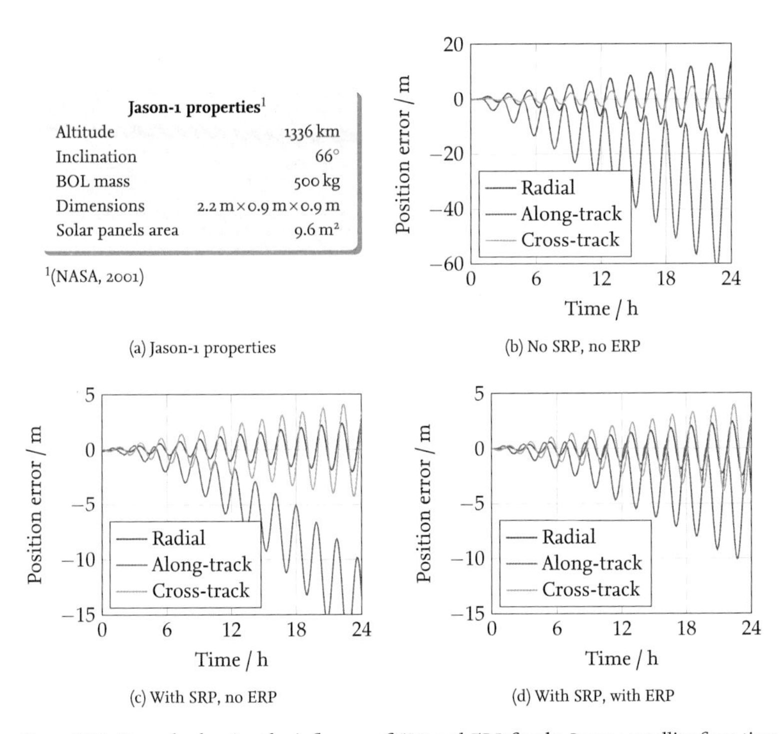

(a) Jason-1 properties

(b) No SRP, no ERP

(c) With SRP, no ERP

(d) With SRP, with ERP

Figure 2.14.: Example showing the influence of SRP and ERP for the Jason-1 satellite for a time span of 24 h.

(Figure 2.14b), the along-track error after a day of propagation is on the order of 60 m, while the radial error increases to more than 10 m. Including SRP in Figure 2.14c, the along-track error decreases significantly to about 20 m, while the radial and cross-track components are below 5 m. Finally, if ERP is considered (Figure 2.14d), the along-track component is further reduced to 10 m, while radial and cross-track errors remain at the same level.

3

Estimation techniques

3.1. Uncertainty propagation

An essential part of orbital information in an operationally employed satellite catalogue are the uncertainties associated with the state vectors. While the orbit determination process provides the fit residuals as a measure of uncertainty at orbit epoch, the propagation of those errors is crucial for the services provided by an SST system, especially for the collision avoidance service.

The problem of uncertainty propagation can be approached in several different ways, including:

State transition matrix integration With the system dynamics being linearised with respect to a reference trajectory, it is possible to work with the state transition matrix, a concept very familiar from control theory for linear systems (Maybeck, 1979).

Gaussian Mixture Model The non-linear propagation of the uncertainties can be accomplished via the combination of probability density functions for state variables and the computation of their evolution (Giza et al., 2009).

Expansion techniques The time evolution of stochastic processes can be described via polynomial or Taylor expansions, a concept introduced by Wiener (1938). Those methods might be either *intrusive*, performing the propagation based on differential algebra (Berz, 1999; Di Lizia et al., 2008; Armellin et al., 2010); or *non-intrusive*, where the system dynamics are replaced by an expansion of the uncertainties in truncated series (Jones et al., 2013; Riccardi et al., 2015).

Monte Carlo sampling Being also non-intrusive, the Monte Carlo method is used to propagate a large number of sample points within given uncertainty bounds. It considers the full dynamics and reconstructs the variances and covariances from the propagated samples. An example for the sequential orbit estimation is the *Sigma Point Filtering* (Lee and Alfriend, 2007).

The state transition matrix will be an important element in the GAMBIT method, based on finding a least-squares solution, hence it will be introduced in the following section. Monte Carlo sampling can be used to verify the correct implementation of the variational equations and the integration process to some extent: it is important to keep in mind that there will be differences arising from the non-linear state propagation when compared to a linearised model.

The propagation of the uncertainties, which are part of the covariance matrix, are important for sequential filters, like the Extended Kalman Filter (EKF). The time update of the Kalman filter requires the assessment of the process noise.

Both Nazarenko (2010) and Wright (1981) point out that there is a problem with modelling it as white noise, as basically all relevant perturbation models come with time-correlated errors. Although computationally expensive, a proper modelling of the non-Gaussian noise can be beneficial in the orbit determination solution, a concept introduced in Section 3.4.4. Although neglected in many applications, accounting for process noise, especially in sequential filters, significantly reduces the effect of *filter smugness* (Vallado and McClain, 2013): The covariance matrix may diverge and become very small, increasing the risk of the filter starting to neglect any new observations and thus not incorporating new information into the orbit determination process any longer.

As the integration of the differential equations in the formulation of the covariance matrix integration are linked to the integration of the state vector, Section 3.4.5 provides an example for a possible implementation as used in NEPTUNE .

3.2. State space representation

The uncertainty propagation problem is typically approached by using a state space representation. For a very detailed description of the associated concepts, referencing the textbook of Tapley et al. (2004) is highly recommended, which also served as the basis for the following derivation.

The n-dimensional state vector $\mathbf{x}(t)$ of a satellite can be defined as

$$\mathbf{x}(t) = \begin{bmatrix} \mathbf{r}(t) \\ \mathbf{v}(t) \\ \mathbf{c} \end{bmatrix}, \tag{3.1}$$

with $\mathbf{r}(t)$and $\mathbf{v}(t)$being the radius and velocity vector, respectively, while $\mathbf{c}$ is a multi-dimensional vector of constant parameters to be solved for in the orbit determination process. Examples for those parameters are the drag and SRP coefficients. Due to the non-linear nature of orbital motion, which is directly seen from Equation 2.1, the conversion of the second-order differential equations of motion into a vector of first-order differential equations is written as

$$\frac{d}{dt}\mathbf{x}(t) = \mathbf{f}(\mathbf{x}, t). \tag{3.2}$$

As the state, in general, cannot be observed directly, an additional equation, the *measurement model* (Maybeck, 1979), is required to map the state vector to the observations:

$$\mathbf{y}(t) = \mathbf{g}(\mathbf{x}, t) + \epsilon, \tag{3.3}$$

here, ϵ is the error in the observation. With $\mathbf{y}(t)$ being p-dimensional and, in general, $p < n$ (number of individually observed quantities is smaller than the number of state vector components), the number of observations m in the orbit determination context guarantees that $m \cdot p \gg n$.

A linearisation is obtained by introducing a reference trajectory, denoted by an index r, and to introduce the deviations as

$$\Delta \mathbf{x}(t) = \mathbf{x}(t) - \mathbf{x}_r(t), \tag{3.4}$$

$$\Delta \mathbf{y}(t) = \mathbf{y}(t) - \mathbf{y}_r(t), \tag{3.5}$$

which are assumed to be small - a concept familiar to Encke's method for perturbed orbits. Expanding Equation 3.2 and Equation 3.3 about the reference state gives

$$\dot{\mathbf{x}}(t) = \mathbf{f}(\mathbf{x}_r, t) + \left(\frac{\partial \mathbf{f}(\mathbf{x}, t)}{\partial \mathbf{x}(t)}\right)_r (\mathbf{x}(t) - \mathbf{x}_r(t)) + \mathcal{O}\left(\Delta \mathbf{x}(t)^2\right), \tag{3.6}$$

$$\mathbf{y}(t) = \mathbf{g}(\mathbf{x}_r, t) + \left(\frac{\partial \mathbf{g}(\mathbf{x}, t)}{\partial \mathbf{x}(t)}\right)_r (\mathbf{x}(t) - \mathbf{x}_r(t)) + \mathcal{O}\left(\Delta \mathbf{x}(t)^2\right) + \epsilon, \tag{3.7}$$

or, combined with Equation 3.4 and Equation 3.5 and neglecting higher-order terms:

$$\Delta \dot{\mathbf{x}}(t) = \left(\frac{\partial \dot{\mathbf{x}}(t)}{\partial \mathbf{x}(t)}\right)_r \Delta \mathbf{x}(t), \tag{3.8}$$

$$\Delta \mathbf{y}(t) = \left(\frac{\partial \mathbf{g}(\mathbf{x}, t)}{\partial \mathbf{x}(t)}\right)_r \Delta \mathbf{x}(t) + \epsilon. \tag{3.9}$$

Defining the system matrix $\mathbf{A}(t)$ as

$$\mathbf{A}(t) \equiv \left(\frac{\partial \dot{\mathbf{x}}(t)}{\partial \mathbf{x}(t)}\right)_r \tag{3.10}$$

and the output matrix as

$$\mathbf{H}(t) \equiv \left(\frac{\partial \mathbf{g}(\mathbf{x}, t)}{\partial \mathbf{x}(t)}\right)_r \tag{3.11}$$

the homogeneous part of the state space representation is obtained for the linearized system:

$$\Delta \dot{\mathbf{x}}(t) = \mathbf{A}(t) \Delta \mathbf{x}(t), \tag{3.12}$$

$$\Delta \mathbf{y}(t) = \mathbf{H}(t) \Delta \mathbf{x}(t) + \epsilon. \tag{3.13}$$

The general solution for Equation 3.12 is:

$$\Delta \mathbf{x}(t) = \frac{\partial \Delta \mathbf{x}(t)}{\partial \Delta \mathbf{x}_0} \Delta \mathbf{x}_0 \tag{3.14}$$

$$= \frac{\partial (\mathbf{x}(t) - \mathbf{x}_r(t))}{\partial (\mathbf{x}_0 - \mathbf{x}_{r,0})} \Delta \mathbf{x}_0 \tag{3.15}$$

$$= \frac{\partial \mathbf{x}(t)}{\partial \mathbf{x}_0} \Delta \mathbf{x}_0 \tag{3.16}$$

given the initial condition $\mathbf{x}(t_0) = \mathbf{x}_0$ and noting that the reference state is constant in the partial derivative in Equation 3.15. Now the state error transition matrix $\mathbf{\Phi}$ can be introduced, which translates the state error $\Delta\mathbf{x}(t)$ from t_0 to t:

$$\mathbf{\Phi}(t, t_0) \equiv \frac{\partial \mathbf{x}(t)}{\partial \mathbf{x}_0} \tag{3.17}$$

Differentiating Equation 3.16 provides

$$\Delta\dot{\mathbf{x}}(t) = \dot{\mathbf{\Phi}}(t, t_0)\,\Delta\mathbf{x}_0, \tag{3.18}$$

which, together with Equation 3.16, can be substituted into Equation 3.12:

$$\dot{\mathbf{\Phi}}(t, t_0)\,\Delta\mathbf{x}_0 = \mathbf{A}(t)\,\mathbf{\Phi}(t, t_0)\,\Delta\mathbf{x}_0, \tag{3.19}$$

finally providing a differential equation for the state transition matrix[1], which will later be integrated numerically for the propagation of the covariance matrix:

$$\dot{\mathbf{\Phi}}(t, t_0) = \mathbf{A}(t)\,\mathbf{\Phi}(t, t_0) \tag{3.20}$$

The main advantage of solving for the state transition matrix using Equation 3.20 over the direct solution of Equation 3.12 is that the state transition matrix allows for a simple formulation of the covariance matrix propagation and the determination of the best estimate of the state vector (Tapley et al., 2004).

The influence of system or process noise, which is characterised by the unmodelled accelerations, is provided by another term leading to the general formulation of the state space representation:

$$\Delta\dot{\mathbf{x}}(t) = \mathbf{A}(t)\,\Delta\mathbf{x}(t) + \mathbf{B}(t)\,\mathbf{u}(t), \tag{3.21}$$

here, $\mathbf{u}(t)$ is the process noise, while $\mathbf{B}(t)$ is the input matrix, which converts the unmodelled accelerations into the quantities of the state vector.

A particular solution to Equation 3.21 can be found via the *variation of constants* method, starting with the product of the state transition matrix $\mathbf{\Phi}(t, t_0)$ and an unknown function $\mathbf{C}(t)$:

$$\Delta\mathbf{x}(t) = \mathbf{\Phi}(t, t_0)\,\mathbf{C}(t). \tag{3.22}$$

Differentiating:

$$\Delta\dot{\mathbf{x}}(t) = \dot{\mathbf{\Phi}}(t, t_0)\,\mathbf{C}(t) + \mathbf{\Phi}(t, t_0)\,\dot{\mathbf{C}}(t), \tag{3.23}$$

and substituting into Equation 3.21 provides

$$\dot{\mathbf{\Phi}}(t, t_0)\,\mathbf{C}(t) + \mathbf{\Phi}(t, t_0)\,\dot{\mathbf{C}}(t) = \mathbf{A}(t)\,\Delta\mathbf{x}(t) + \mathbf{B}(t)\,\mathbf{u}(t). \tag{3.24}$$

This equation can be re-written by substituting the already known relationships for the time derivative of the state error transition matrix, Equation 3.20, and Equation 3.22:

$$\mathbf{A}(t)\,\mathbf{\Phi}(t, t_0)\,\mathbf{C}(t) + \mathbf{\Phi}(t, t_0)\,\dot{\mathbf{C}}(t) = \mathbf{A}(t)\,\mathbf{\Phi}(t, t_0)\,\mathbf{C}(t) + \mathbf{B}(t)\,\mathbf{u}(t), \tag{3.25}$$

[1] In fact, for the linearised system, it is also referred to as the *state error transition matrix*.

which results in

$$\mathbf{\Phi}(t,t_0)\,\dot{\mathbf{C}}(t) = \mathbf{B}(t)\,\mathbf{u}(t). \tag{3.26}$$

The solution is now obtained by integration:

$$\mathbf{C}(t) = \mathbf{C}_0 + \int_{t_0}^{t} \mathbf{\Phi}^{-1}(\xi,t_0)\,\mathbf{B}(\xi)\,\mathbf{u}(\xi)\,d\xi. \tag{3.27}$$

Substituting the result for $\mathbf{C}(t)$ from Equation 3.27 into Equation 3.22 results in:

$$\Delta\mathbf{x}(t) = \mathbf{\Phi}(t,t_0)\,\mathbf{C}_0 + \int_{t_0}^{t} \mathbf{\Phi}(t,t_0)\,\mathbf{\Phi}^{-1}(\xi,t_0)\,\mathbf{B}(\xi)\,\mathbf{u}(\xi)\,d\xi, \tag{3.28}$$

which can be simplified using the following properties of the state transition matrix:

$$\mathbf{\Phi}(t,t_0)\,\mathbf{\Phi}^{-1}(\xi,t_0) = \mathbf{\Phi}(t,t_0)\,\mathbf{\Phi}(t_0,\xi) = \mathbf{\Phi}(t,\xi), \tag{3.29}$$

as well as the initial condition $\mathbf{C}_0 = \mathbf{x}_0$ to finally provide the general solution for the inhomogeneous Equation 3.21:

$$\Delta\mathbf{x}(t) = \mathbf{\Phi}(t,t_0)\,\Delta\mathbf{x}_0 + \int_{t_0}^{t} \mathbf{\Phi}(t,\xi)\,\mathbf{B}(\xi)\,\mathbf{u}(\xi)\,d\xi. \tag{3.30}$$

The result in Equation 3.30 is also referred to as the *matrix superposition integral* (Gelb, 1974), where the second term describes how an input (here: process noise $\mathbf{u}$) at a time ξ translates into a state vector change at time t.

3.3. The covariance matrix

The elements of the state vector are a result of the orbit determination process. As such, the state vector components are always associated with uncertainties and can be assumed as random variables.

The matrix containing the variances of the elements on its diagonal and the covariances of the ith and j^{th} element in row i and column j, is called the *covariance matrix*. It is defined as:

$$\mathbf{P} = E\left[(\mathbf{x} - E[\mathbf{x}])\,(\mathbf{x} - E[\mathbf{x}])^T\right]. \tag{3.31}$$

3.4. Differential correction

The GAMBIT method, which has been developed in this thesis and will be presented in Chapter 4, is based on an algorithm called *Differential Correction*: An initial estimate for the state vector is being iteratively refined in a *batch least squares process* for a given set of new observations. Therefore, the theoretical background will be provided in Section 3.4.1.

Due to the properties of the cost function occurring in the least squares process for typical orbits, convergence is not always guaranteed (see Chapter 4 for an example). An optimization technique, known as the Levenberg-Marquardt algorithm (LM), makes the process more robust by controlling the convergence process. It will be introduced in Section 3.4.2.

3.4.1. Non-linear least squares

For the determination of a satellite's orbit, a set of measurements of the trajectory at discrete times is available, which have to be processed to obtain the state vector of that object for a given epoch. This leads to a *non-linear least squares* problem due to the non-linear nature of the equations of motion (Equation 3.2). As the methods presented in Chapter 4 are processing a *batch* of data (or observations) in order to obtain an estimate, they are referred to as *batch least square* methods, as opposed to sequential methods, e.g. the Kalman filter.

The problem consists in minimising the cost function or performance index $J(\mathbf{x})$ (Gelb, 1974) by selecting an appropriate state vector $\mathbf{x}$, according to:

$$J(\mathbf{x}) = \frac{1}{2} \cdot \epsilon^T \cdot \mathbf{W} \cdot \epsilon = \frac{1}{2} \cdot \sum_{i=1}^{m} w_{ii} \cdot \epsilon_i^2, \tag{3.32}$$

here, the observation error ϵ, or *observation residual*, is obtained as the difference between the observation and the computed measurement. Assuming a diagonal weight matrix $\mathbf{W}$, its components w_{ii} allow for individual observations being weighted according to their quality and type.

For the linearised measurement model from Equation 3.13 the observation residuals are given as:

$$\epsilon = \Delta\mathbf{y}(t) - \mathbf{H}(t) \cdot \Delta\mathbf{x}(t). \tag{3.33}$$

Non-linear optimisation problems are solved iteratively, so that the best estimate of the state, $\hat{\mathbf{x}}$, will result from a series of different state vectors $(\mathbf{x}_0, \mathbf{x}_1, \ldots)$ converging towards that solution. For a minimised cost function, one can thus write:

$$J(\Delta\hat{\mathbf{x}}) = \frac{1}{2} \cdot (\Delta\mathbf{y} - \mathbf{H} \cdot \Delta\hat{\mathbf{x}})^T \cdot \mathbf{W} \cdot (\Delta\mathbf{y} - \mathbf{H} \cdot \Delta\hat{\mathbf{x}}) \tag{3.34}$$

A minimum of the scalar is obtained, when the derivative with respect to the best estimate is zero (or null vector) and the Hessian of J is positive definite (Gelb, 1974; Tapley et al., 2004):

$$\frac{\partial J}{\partial \hat{\mathbf{x}}} = \mathbf{0} \quad \text{and} \quad \Delta\hat{\mathbf{x}}^T \cdot \left(\frac{\partial^2 J}{\partial \hat{\mathbf{x}}^2}\right) \cdot \Delta\hat{\mathbf{x}} > 0, \quad \forall \Delta\hat{\mathbf{x}} \neq 0$$

After differentiating and setting to zero, the following result is obtained:

$$\mathbf{H}^T \cdot \mathbf{W} \cdot \mathbf{H} \cdot \Delta\hat{\mathbf{x}} = \mathbf{H}^T \cdot \mathbf{W} \cdot \Delta\mathbf{y} \tag{3.35}$$

The second derivative of the cost function is

$$\frac{\partial^2 J}{\partial \hat{\mathbf{x}}^2} = \mathbf{H}^T \cdot \mathbf{W} \cdot \mathbf{H}, \tag{3.36}$$

a symmetric $n \times n$ matrix, which will be positive definite, if $\mathbf{H}$ is full rank. This is guaranteed as long as the number of linearly independent observations is greater than the dimension of the state vector (Tapley et al., 2004). As positive definite matrices are always invertible [2], the result for the best estimate can be obtained via

$$\Delta\hat{\mathbf{x}} = \left(\mathbf{H}^T \cdot \mathbf{W} \cdot \mathbf{H}\right)^{-1} \cdot \mathbf{H}^T \cdot \mathbf{W} \cdot \Delta\mathbf{y}. \tag{3.37}$$

Solving iteratively for the best estimate

$$\hat{\mathbf{x}}_{k+1} = \hat{\mathbf{x}}_k + \Delta\hat{\mathbf{x}}_k, \tag{3.38}$$

corresponds to the *Gauss-Newton method*, which is of quadratic convergence, when $\hat{\mathbf{x}}_k$ is close to the local minimum (Madsen et al., 2004):

$$|\epsilon_{k+1}| = \mathcal{O}\left(|\epsilon_k|^2\right), \text{ when } |\epsilon_k| \text{ is small.} \tag{3.39}$$

In general, however, it is of linear convergence (Madsen et al., 2004):

$$|\epsilon_{k+1}| \leqslant a \cdot |\epsilon_k|, \text{ when } |\epsilon_k| \text{ is small; } 0 < a < 1. \tag{3.40}$$

It has to be noted, however, that convergence for the Gauss-Newton method is not guaranteed. Moreover, there is not even a guarantee for local convergence, which is clearly the case for linear least squares via *Newton's method.*

3.4.2. Levenberg-Marquardt algorithm

The Levenberg-Marquardt method (LM) is also known as a *damped* or *trusted region* approach to obtain the least-squares solution. It is considered to be more robust than the Gauss-Newton method presented in the previous section, which means that it is very likely to converge even under unfavourable initial conditions.

Writing, for convenience, the step of the LM method as $\mathbf{h}_{LM} = \Delta\hat{\mathbf{x}}$, Levenberg (1944) introduced the following modification to Equation 3.37:

$$\mathbf{h}_{LM} = \left(\mathbf{H}^T \cdot \mathbf{W} \cdot \mathbf{H} + \lambda_{LM} \cdot \mathbf{I}\right)^{-1} \cdot \mathbf{H}^T \cdot \mathbf{W} \cdot \Delta\mathbf{y}. \tag{3.41}$$

The damping parameter λ_{LM} has several effects:

- For $\lambda_{LM} > 0$, the inverse in Equation 3.41 will be positive definite, ensuring that the update step will be in descent direction (Madsen et al., 2004).
- For λ_{LM} being very large, the step is in *steepest descent* direction (negative gradient):

$$\mathbf{h}_{LM} \simeq \frac{1}{\lambda_{LM}} \cdot \mathbf{H}^T \cdot \mathbf{W} \cdot \Delta\mathbf{y}. \tag{3.42}$$

  which means that a short step in descent direction will be made. This is the desired behaviour, if the current iterate is far from the solution (Madsen et al., 2004).

[2] If the positive definite matrix $\mathbf{A}$ was not invertible, there must be a non-zero $\mathbf{x}$ such that $\mathbf{A} \cdot \mathbf{x} = \mathbf{0}$. This means that $\mathbf{x}^T \cdot \mathbf{A} \cdot \mathbf{x} = 0$, contradicting the assumption that $\mathbf{A}$ is positive definite.

- For very small λ_{LM}, the step will be similar to the one obtained via the Gauss-Newton method (Equation 3.37). This is beneficial for the final (almost quadratic) convergence.

It can thus be stated, that the LM interpolates between two different methods, the steepest descent and the Gauss-Newton.

The choice of the initial damping parameter upon initialisation of the method, $\lambda_{LM,0}$, can be related to the Fisher information $\mathbf{Z} = \mathbf{H}^T \cdot \mathbf{W} \cdot \mathbf{H}$ (Madsen et al., 2004):

$$\lambda_{LM,0} = \chi_0 \max_i \left(z_{ii}^{(0)} \right), \tag{3.43}$$

where χ_0 is a tuning parameter of the algorithm and z_{ii} are the diagonal elements of the Fisher information matrix $\mathbf{Z}$. Madsen et al. (2004) provide a scheme for the update of the damping parameter λ_{LM} during the iteration. It is controlled by the gain ratio ρ_{LM}(Madsen et al., 2004):

$$\rho_{LM} = \frac{J(\hat{\mathbf{x}}_k) - J(\hat{\mathbf{x}}_k + \mathbf{h}_{LM})}{L(\mathbf{0}) - L(\mathbf{h}_{LM})}, \tag{3.44}$$

assuming that there is a model L of the behaviour of J in the neighbourhood of the current iterate $\hat{\mathbf{x}}_k$, which can be a Taylor expansion of J around $\hat{\mathbf{x}}_k$:

$$J(\hat{\mathbf{x}}_k + \mathbf{h}_{LM}) \simeq L(\mathbf{h}_{LM}) \equiv J(\hat{\mathbf{x}}) + \mathbf{h}_{LM}^T \cdot \mathbf{H}^T \cdot \mathbf{W} \cdot \Delta\mathbf{y} + \frac{1}{2} \cdot \mathbf{h}_{LM}^T \cdot \mathbf{H}^T \cdot \mathbf{W} \cdot \mathbf{H} \cdot \mathbf{h}_{LM} \tag{3.45}$$

For the denominator in Equation 3.44 one obtains:

$$L(\mathbf{0}) - L(\mathbf{h}_{LM}) = \frac{1}{2} \cdot \mathbf{h}_{LM}^T \cdot \left(\lambda_{LM} \cdot \mathbf{h}_{LM} - \mathbf{H}^T \cdot \mathbf{W} \cdot \Delta\mathbf{y} \right). \tag{3.46}$$

It is important to note that Equation 3.46 will be always positive, as both, $\mathbf{h}_{LM}^T \mathbf{h}_{LM}$ and, considering Equation 3.37, also $-\mathbf{h}_{LM}^T \cdot \mathbf{H}^T \cdot \mathbf{W} \cdot \Delta\mathbf{y}$ are positive. In general, one would expect the predicted decrease in the cost function (denominator in Equation 3.44) to be higher than the actual decrease (nominator in Equation 3.44), which is due to the damping for the latter. Hence, higher values for ρ_{LM} mean that there is a good approximation by the model or, even better, the decrease in the cost function has been even larger than predicted, so that for the next step, the damping via λ_{LM} can be reduced. On the other hand, small or even negative values for ρ_{LM} indicate to increase the damping.

Nielsen (1999) provides an update scheme for λ_{LM} based on the gain ratio, which is done without any additional evaluations of the cost function (Algorithm 3.1). The damping parameter will be quickly increased by doubling, each time a step fails. For accepted steps, the maximum reduction of λ_{LM} is dividing it by three. By selecting these specific values (2 and $\frac{1}{3}$), it cannot happen, that λ_{LM} will bounce back and forth between two values. The bridging function connecting those two values allows for a smoother performance and even a faster convergence, as Nielsen (1999) points out.

Algorithm 3.1: Update of the Levenberg-Marquardt damping parameter.

if $\rho_{LM} > 0$ **then**
$\lambda_{LM} := \lambda_{LM} \cdot \max\left(\frac{1}{3}, 1 - (2\rho_{LM} - 1)^3\right)$;
$\nu := 2$;
else
$\lambda_{LM} := \lambda_{LM} \cdot \nu$;
$\nu := 2 \cdot \nu$;
end

3.4.3. Propagation using the State Transition Matrix

Using the linearised system in the state space representation, as introduced in Section 3.2, the expected value of the state vector is equal to the reference orbit:

$$E[\mathbf{x}] = \mathbf{x}_r(t). \tag{3.47}$$

Substituting this result and Equation 3.4 into Equation 3.31, the covariance matrix at a time t can be written as:

$$\mathbf{P}(t) = E\left[\Delta\mathbf{x}(t)\,\Delta\mathbf{x}(t)^T\right]. \tag{3.48}$$

With the matrix superposition integral from Equation 3.30, one obtains:

$$\mathbf{P}(t) = \mathbf{\Phi}(t,t_0)\,\mathbf{P}_0\mathbf{\Phi}(t,t_0)^T + \mathbf{Q}_{xu}(t) + \mathbf{Q}_{xu}(t)^T + \mathbf{Q}_{uu}(t), \tag{3.49}$$

where the first term is the time update of the covariance matrix from t_0 to t, with

$$\mathbf{P}_0 = \mathbf{P}(t_0) = E\left[\Delta\mathbf{x}(t_0)\,\Delta\mathbf{x}(t_0)^T\right], \tag{3.50}$$

the matrix $\mathbf{Q}_{xu}$ is the cross-correlation of the state vector $\mathbf{x}$ at t_0 and the process noise $\mathbf{u}$ at time t:

$$\mathbf{Q}_{xu}(t) = \mathbf{\Phi}(t,t_0)\cdot\int_{t_0}^{t} E\left[\Delta\mathbf{x}(t_0)\cdot\mathbf{u}(\eta)^T\right]\cdot\mathbf{B}(\eta)\cdot\mathbf{\Phi}(t,\eta)\,d\eta, \tag{3.51}$$

and the last term is the auto-correlation function of the noise:

$$\mathbf{Q}_{uu}(t) = \int_{t_0}^{t}\int_{t_0}^{t} \mathbf{\Phi}(t,\xi)\cdot\mathbf{B}(\xi)\cdot E\left[\mathbf{u}(\xi)\cdot\mathbf{u}(\eta)^T\right]\mathbf{B}(\eta)^T\cdot\mathbf{\Phi}(t,\eta)^T\,d\xi d\eta. \tag{3.52}$$

The state transition matrix used for the time update of the covariance matrix can be obtained via a numerical integration of Equation 3.20. An implementation example for the propagator Neptune will be briefly outlined in Section 3.4.5.

While many applications implement the time update of the covariance matrix assuming the state at t_0 and the noise as uncorrelated and, furthermore, no time correlation for the noise, some authors (Nazarenko, 2010; Wright et al., 2008) have emphasized that especially for orbit propagation, those assumptions do not hold. Therefore, the next section shall provide a short introduction into modelling process noise accounting for noise and state vector cross-correlation as well as for noise auto-correlation.

3.4.4. Process noise

The propagation of a perturbed orbit is based on force models which are manifold, including: uncertainties in the coefficients of the geopotential; the representation of the geopotential through a truncated series; the stochastic nature of solar and geomagnetic activity; uncertainty in the ballistic coefficient; simplifications in modelling the optical properties; etc.

Most of the above mentioned effects may be described by a stochastic process. For many systems, such a process is well approximated by Gaussian white noise and is referred to as State Noise Compensation (SNC). A more sophisticated modelling, where process noise parameters are included in the orbit determination as solve-for parameters, is referred to as Dynamic Model Compensation (DMC). Both methods are described in detail by Tapley et al. (2004).

However, as Nazarenko (2010) points out, the analysis of the geopotential model reveals that the resulting errors are time-correlated and therefore non-Gaussian, a property generally denoted as *coloured noise*.

Both, the auto-correlation and cross-correlation functions, being part of Equation 3.52 and Equation 3.51, respectively, are provided by Nazarenko (2010) for the geopotential. While accelerations due to the geopotential, in theory, can be computed to an arbitrary accuracy via Equation 2.26, in practice the series is truncated for computational reasons, thereby introducing errors on the order of the omitted coefficients. Moreover, even the considered spherical harmonics up to degree n and order m introduce errors through the errors in the determination of the Stokes coefficients.

Auto-correlation function of the geopotential model errors

For the geopotential, the process noise vector $\mathbf{u}$ results from the errors in the acceleration obtained from the model and can be conveniently defined in the orbit reference frame:

$$\mathbf{u}(t) = (\Delta f_U, \Delta f_V, \Delta f_W)^T, \tag{3.53}$$

here, Δf_U is the error in the force model in radial, Δf_V in along-track and Δf_W in orbit normal direction, respectively.

The auto-correlation function can then be written as:

$$E\left[\mathbf{u}\mathbf{u}^T\right] = E\begin{bmatrix} \Delta f_U^2 & \Delta f_U \Delta f_V & \Delta f_U \Delta f_W \\ \Delta f_V \Delta f_U & \Delta f_V^2 & \Delta f_V \Delta f_W \\ \Delta f_W \Delta f_U & \Delta f_W \Delta f_V & \Delta f_W^2 \end{bmatrix} \approx \begin{pmatrix} 1 & 0 & 0 \\ 0 & \frac{1}{2} & 0 \\ 0 & 0 & \frac{1}{2} \end{pmatrix} K_U. \tag{3.54}$$

In the above equation, the first matrix expression contains the required derivatives. Nazarenko (2010) shows, that all non-diagonal terms are zero and provides the variances on the diagonal. The outer right expression in Equation 3.54 is the final result after introducing a simplification due to the similarity between the three components. The function

$K(r,\psi)$ is the auto-correlation function of two points separated by a central angle ψ on a sphere of radius **r** (Nazarenko, 2010):

$$K(r,\psi) = \left(\frac{\mu}{r^2}\right)^2 \sum_{n=2}^{\infty} \left(\frac{R_\oplus}{r}\right)^{2n} \frac{(n+1)^2(2n+1)}{2} \left(\Delta\overline{C}^2_{n,m} + \Delta\overline{S}^2_{n,m}\right) P_{n,0}(\cos\psi), \quad (3.55)$$

where the angular distance can be computed as a function of the geocentric latitude $\phi_{gc}$and longitude λ:

$$\cos\psi = \sin\phi_{gc,1}\sin\phi_{gc,2} + \cos\phi_{gc,1}\cos\phi_{gc,2}\cos(\lambda_1 - \lambda_2). \quad (3.56)$$

The sum in Equation 3.55 is with respect to the degree of the geopotential. The quantities $\Delta\overline{C}_{n,m}$ and $\Delta\overline{S}_{n,m}$ for a given degree n are combined according to (see also Equation 2.27 and Equation 2.28):

$$\left(\Delta\overline{C}^2_{n,m} + \Delta\overline{S}^2_{n,m}\right) = \begin{cases} \frac{2}{2n+1} \sum_{m=0}^{n} \left(\sigma^2\left(\overline{C}_{n,m}\right) + \sigma^2\left(\overline{S}_{n,m}\right)\right), & \text{if } n \leqslant n_{max} \\ \frac{10^{-5}}{n^2} & \text{if } n > n_{max} \end{cases} \quad (3.57)$$

This means that the standard deviations of the harmonic coefficients are summed up to a sufficiently high maximum degree of the geopotential considered in the propagation ($n \leqslant n_{max}$), while the harmonic coefficients are used for $n > n_{max}$. As Nazarenko (2010) points out in his analysis based on the EGM96 model, the computation of the series in Equation 3.55 can be carried out in two ways:

- For n_{max} being small, the series will be truncated early, so that the error due to omission will be dominated by larger terms, with the influence of terms of increasing order quickly declining. This means that the series with respect to the geopotential degree would be evaluated up to $n \approx 30\ldots40$.
- For the geopotential being considered to a high degree for high precision computations, the truncation sets in late and the amount of terms that are non-negligible will be higher than in the case above. In order to reduce computational burden, Nazarenko (2010) recommends to use Kaula's rule (Kaula, 2000) for an efficient evaluation (see Equation 2.28).

In Figure 3.1 the *normalized* auto-correlation function is shown as a function of the angular distance for different orbit altitudes and two different values for n_{max}. The normalization is with respect to the value at $\psi = 0$:

$$k(r,\psi) = \frac{K(r,\psi)}{K(r,\psi=0)} \quad (3.58)$$

It can be seen, that the correlation is non-negligible and has higher values for increasing orbital altitude and decreasing degree of the geopotential.

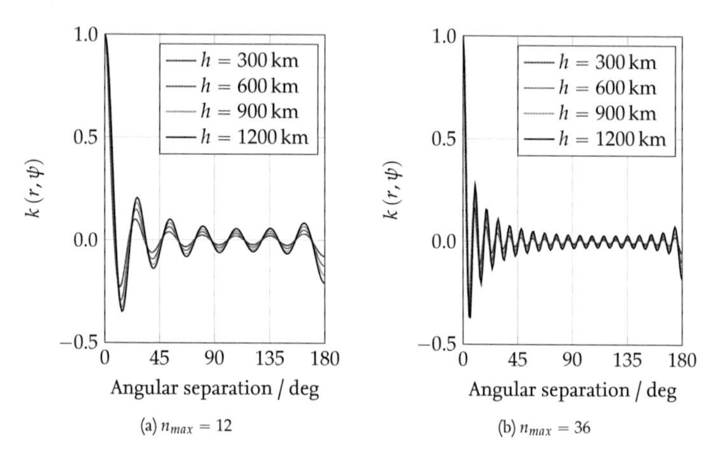

Figure 3.1.: Normalized auto-correlation function used to compute noise contributions from geopotential model errors. The shown results are for geopotential degrees $n = 12$ (left) and $n = 36$ (right), for different altitudes.

The normalised auto-correlation function $k(r, \psi)$ is shown in Figure 3.2 for the argument of true latitude $u(\xi)$ of the orbit under consideration (i.e. at epoch t_0) against the argument of true latitude $u(\eta)$ of the same orbit and subsequent revolutions. It can be seen, that for $u(\eta) = u(\xi)$ the normalised function value is $k = 1$, as expected. However, it can also be noted, that for $u(\eta) \neq u(\xi)$, the function value is not always zero - which would be the case for the Gaussian white noise assumption. One also finds that normalised auto-correlation function is $k = 1$ at those points, where subsequent groundtracks intersect. For most orbits, this happens twice per revolution, resulting in the elliptic patterns for $u(\eta)$ and $u(\xi)$ in different orbit revolutions. In Figure 3.2c one can see, that for repeating groundtracks, as is the case for every second orbit for GPS navigation satellites, the auto-correlation function will provide significant contributions for the evaluation of the integral in Equation 3.52.

By computing that integral, the matrix $\mathbf{Q}_{uu}(t)$ is finally obtained and is added to the propagated covariance matrix at time t, the latter containing only the modelled time evolution of the uncertainties via the variational equations. Thus, the covariance matrix at each propagated step will be larger than would be the case without noise. A few examples are shown in Figure 3.3 for the standard deviation of the radial position error. An initial standard deviation of 1 m in radial direction was assumed, while all other elements of the covariance matrix were set to zero, in order to have a simplified example. It can be seen that the radial error remains bounded for subsequent revolutions, with a maximum of about 3 m, if one propagates without noise. However, as soon as the matrix $\mathbf{Q}_{uu}(t)$ is considered, the radial error increases significantly, especially if the altitude is lower and the

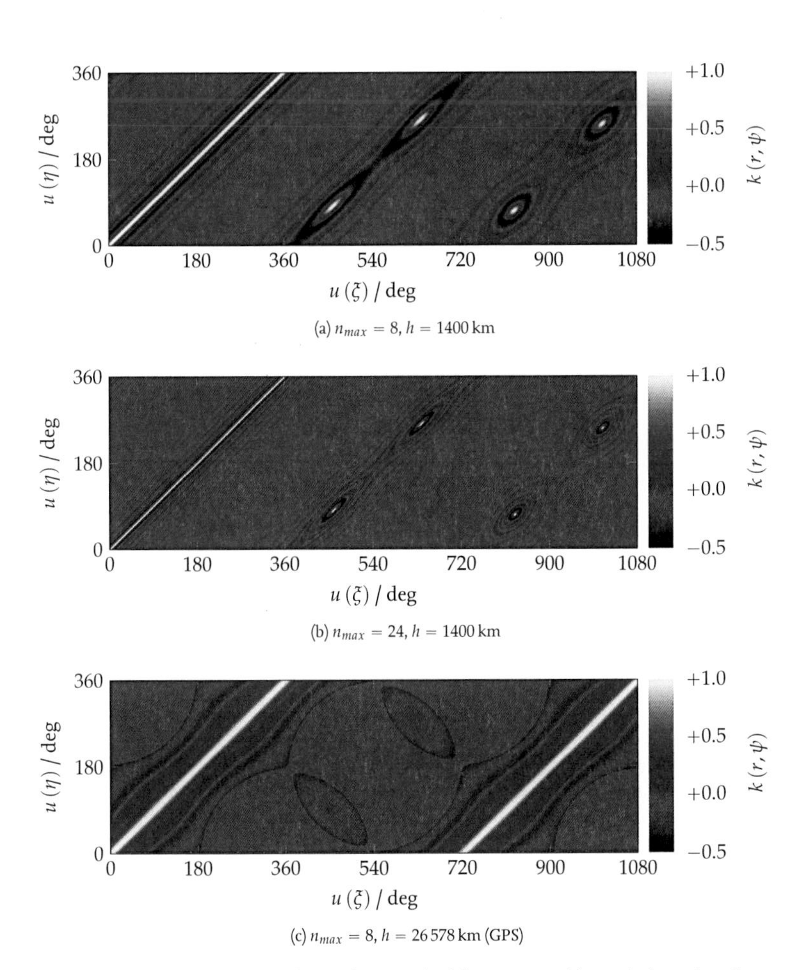

(a) $n_{max} = 8$, $h = 1400\,\text{km}$

(b) $n_{max} = 24$, $h = 1400\,\text{km}$

(c) $n_{max} = 8$, $h = 26\,578\,\text{km}$ (GPS)

Figure 3.2.: Normalised auto-correlation function k of the current orbit revolution, given by argument of true latitude $u(\eta)$, with the current and two subsequent orbits, given by $u(\xi)$. Shown for different altitudes and maximum degree of the geopotential; same inclination of $i = 55°$ for all orbits.

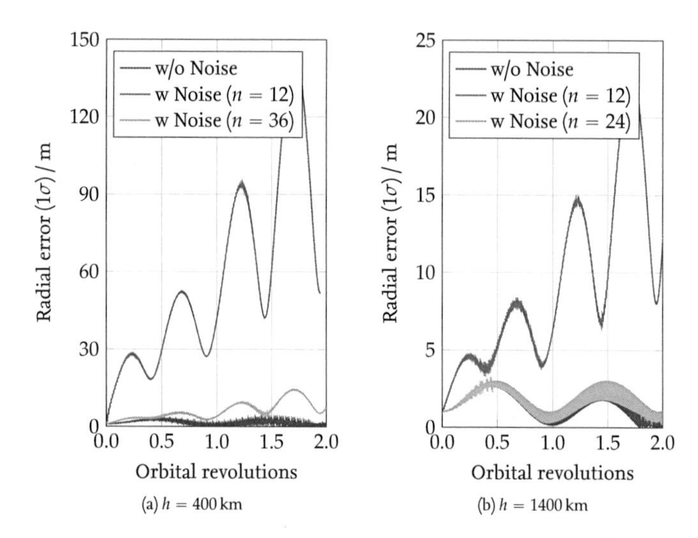

Figure 3.3.: Radial error (1σ) for different altitudes, comparing covariance matrix propagation with and without considering noise for varying degree of the geopotential. An initial radial error of 1 m at t_0 was assumed.

series for the geopotential is truncated for lower degrees. In the example in Figure 3.3b one can also see that the result is almost the same with and without $\mathbf{Q}_{uu}(t)$ for $n = 36$.

Cross-correlation function of the geopotential model errors and the state vector errors at epoch

Nazarenko (2010) also points out that the integral for the cross-correlation between the state vector at epoch and the geopotential model errors is not zero in general and needs to be computed in Equation 3.51 for a more realistic assessment of the associated uncertainties. With

$$E\left[\Delta\mathbf{x}(t_0)\,\mathbf{u}(t)^T\right] \neq 0, \tag{3.59}$$

the evaluation of the cross-correlation integral is analogously to the above scheme for the auto-correlation function. For more details the interested reader is recommended to study the book of Nazarenko (2010).

3.4.5. Integration example: Neptune

To conclude on the propagation part, this section provides an overview on how the different computations for the state vector and the covariance matrix are actually performed. Many authors settle the issue with the remark that the differential equations for the covariance matrix in Equation 3.20 are conveniently solved together with the integration of the

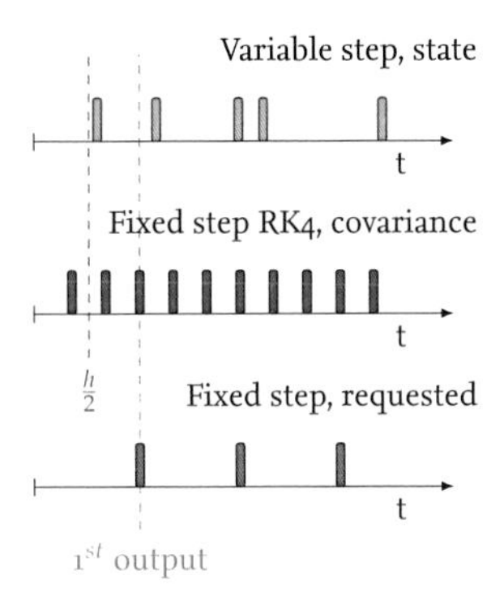

Figure 3.4.: Example for the parallel integration of state vector (top) and the covariance matrix (center), where the latter is shown as a RK4 method. Individual steps are shown on a timeline, including the requested output steps (bottom).

accelerations from the force model for the state vector. However, this is not trivial when a variable-step approach is used for the state vector computation, as done in NEPTUNE (see Section 2.1): the step size is controlled according to the accepted tolerance and solving for the state transition matrix in Equation 3.20 might require a completely different step size.

Therefore, the idea is to separate the propagation of the covariance matrix and the state vector in the essential part, which is the integrator. This comes with another advantage, as the variable-step Störmer-Cowell integrator presented in Section 2.1 has a large overhead compared to other methods: For the (at least) 36 differential equations of the state transition matrix, the force model evaluation (in terms of the variational equations) can be quite fast, as typically only the two-body solution and the oblateness of the Earth (characterised by the J_2 coefficient) have to be considered. Using a complex integrator would thus lead to an unfavourable ratio of integrator overhead to force model computations.

It is therefore convenient to select a self-starting and fixed-step integrator to solve Equation 3.20. An example is shown in Figure 3.4 for the RK4 method, which requires three evaluations of the force model: at t_0, $t_0 + \frac{h}{2}$ and $t_0 + h$, where h is the integration step size. For example, in the two-body case, the derivative of the state transition matrix at t_0, $\dot{\mathbf{\Phi}}(t_0, t_0)$, can be evaluated as:

$$\dot{\mathbf{\Phi}}(t_0, t_0) = \mathbf{A}_{2b}(t_0)\,\mathbf{\Phi}(t_0, t_0) = \mathbf{A}_{2b}(t_0) = \frac{\partial \dot{\mathbf{x}}}{\partial \mathbf{x}} = \begin{pmatrix} 0 & 0 & 0 & 1 & 0 & 0 \\ 0 & 0 & 0 & 0 & 1 & 0 \\ 0 & 0 & 0 & 0 & 0 & 1 \\ \frac{\partial a_x}{\partial r_x} & \frac{\partial a_x}{\partial r_y} & \frac{\partial a_x}{\partial r_z} & 0 & 0 & 0 \\ \frac{\partial a_y}{\partial r_x} & \frac{\partial a_y}{\partial r_y} & \frac{\partial a_y}{\partial r_z} & 0 & 0 & 0 \\ \frac{\partial a_z}{\partial r_x} & \frac{\partial a_z}{\partial r_y} & \frac{\partial a_z}{\partial r_z} & 0 & 0 & 0 \end{pmatrix} \tag{3.60}$$

with

$$\frac{\partial a_x}{\partial r_x} = 3\mu\frac{r_x^2}{r^5} - \frac{\mu}{r^3} \tag{3.61}$$

$$\frac{\partial a_y}{\partial r_y} = 3\mu\frac{r_y^2}{r^5} - \frac{\mu}{r^3} \tag{3.62}$$

$$\frac{\partial a_z}{\partial r_z} = 3\mu\frac{r_z^2}{r^5} - \frac{\mu}{r^3} \tag{3.63}$$

$$\frac{\partial a_x}{\partial r_y} = \frac{\partial a_y}{\partial r_x} = 3\mu\frac{r_x r_y}{r^5} \tag{3.64}$$

$$\frac{\partial a_x}{\partial r_z} = \frac{\partial a_z}{\partial r_x} = 3\mu\frac{r_x r_z}{r^5} \tag{3.65}$$

$$\frac{\partial a_y}{\partial r_z} = \frac{\partial a_z}{\partial r_y} = 3\mu\frac{r_y r_z}{r^5}. \tag{3.66}$$

In the Equations 3.61 - 3.66, the state vector r is the required input quantity. Now, Figure 3.4 shows, how the results from the state vector integration can be beneficially used for a fast integration of the state transition matrix: The variable-step Störmer-Cowell integrator (shown at the top in Figure 3.4) performs its first step from t_0 to t. Assuming that the covariance matrix with a fixed step integration proceeds with a smaller step size, i.e. $h < t - t_0$, the required state vector inputs for the state transition matrix at t_0, $t_0 + \frac{h}{2}$ and $t_0 + h$ are obtained via a fast interpolation (Section 2.1.5), without any force model evaluation. Hence, the integration of the differential equations for the state transition matrix is very fast, even for very small step sizes.

Computation of the matrix $\mathbf{Q}_{uu}$

Besides the derivation of a method to assess coloured noise due to the errors in the geopotential, Nazarenko (2010) also analyses computational issues in-depth. Using non-singular orbit elements (NSO) for the computation of Equation 3.52, it is possible to pre-compute the matrix $\mathbf{Q}_{uu}$ for the entire propagation span and interpolate for the requested output times. In order to be combined with the propagated covariance matrix in cartesian space (CS), a transformation is required:

$$\mathbf{P}_{CS} = \mathbf{J} \cdot \mathbf{P}_{NSO} \cdot \mathbf{J}^T, \tag{3.67}$$

where the Jacobian $\mathbf{J}$ contains the partial derivatives of the radius and velocity vector with respect to the non-singular orbit elements.

4

Methods providing orbit information of predetermined bounded accuracy

With the broad theoretical framework derived in the previous chapters, the methodology to derive orbits of a given accuracy shall be presented in the following.

The first step is to define what is expected from such a method in Section 4.1. The motivation for using such a method in an SST system is once more emphasized by the examination of the currently used technique in the operational SSN to derive TLE in Section 4.2.

Different approaches to obtain orbit information with degraded accuracy, which fulfill the defined goals, shall be presented in Section 4.3.

4.1. Pre-considerations and method requirements

For an operational space object catalog, being the core element of an SST system (see Figure 1.5), it is self-evident to store data and information with the best possible accuracy for each object. The individual services such a system supports rely on this data, but for many of them also less accurate data is fully sufficient. For example, passive optical telescopes may still acquire a satellite for tracking, even if only coarse orbit information like propagated orbits from TLE are used as a priori information.

One can thus imagine a functionality, which takes the most accurate orbit information from the catalogue and, in a still to be defined manner, converts it to a product that is sufficient for the corresponding service. For the entity operating such a system, that functionality would allow for more diversified services. On the other hand, working with less accurate solutions often (depending on the applied methodology) allows for having less data to transfer and to process: for instance, if an object had to be tracked during a night, highly accurate ephemerides would have to be provided in one-minute steps, e.g. via an OEM, while a simple TLE could have the same effect with only two short lines of data.

4.1.1. Accuracy, precision and trueness

The first important point to be addressed is the term *accuracy* itself. What do we mean when we speak of an orbit being *accurate*? Intuitively, this should seem to be a palpable concept for the reader. However, there are also terms like *precision* and *trueness*, which are often synonymously or interchangeably used in literature.

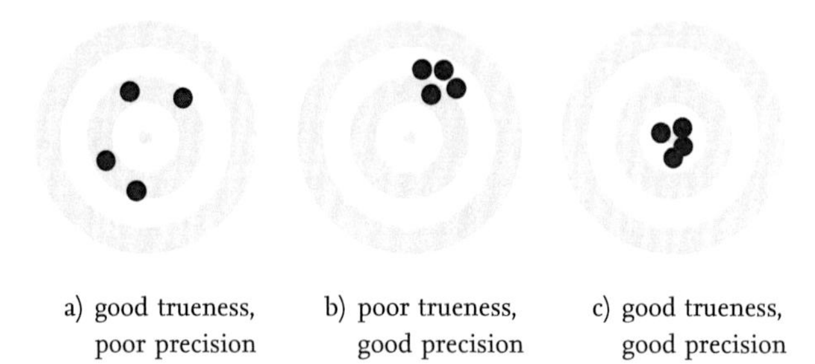

a) good trueness, poor precision b) poor trueness, good precision c) good trueness, good precision

Figure 4.1.: Accuracy, precision and trueness visualised. While accuracy is poor in a) and b), when either precision or trueness are poor, one can consider a solution to be accurate only if it has a good trueness and precision, as shown in c).

The scientific method requires us to have a clear understanding what accuracy means and how it is discriminated against precision or trueness. The ISO 5725-1:1994 (ISO, 1994) defines:

Accuracy closeness of agreement between a test result or measurement result and the true value.

Trueness closeness of agreement between the expectation of a test result or measurement result and a true value.

Precision closeness of agreement between independent test/measurement results obtained under stipulated conditions.

The meaning of these definitions can be easily visualised, as shown in Figure 4.1. According to the ISO definition, *accuracy* is thus a combination of *precision* and *trueness* and can be considered superordinate to the latter two. If either precision or trueness are poor, accuracy will be poor as well. Only if both, precision and trueness are good, one can also speak of a good accuracy.

With the above reasoning in mind, it is thus convenient to refer to a method that provides orbit information of *degraded* or, more neutral, *predetermined accuracy*, irrespective of how the specific method is defined. One can imagine different methods affecting the trueness, the precision or both.

4.1.2. Reference orbit

In the previous section, the term accuracy has been defined by means of a comparison of a test result to a *true value*. The *true value* which is required to assess the accuracy, is not accessible in practice, and needs thus to be replaced by an agreed reference value (ISO, 1994).

The most convenient reference orbit for the SST system is the catalogued orbit of an object, i.e. the most accurate solution available. It is important to note that such a reference will always be a result of a POD process and thus itself subject to varying levels of trueness and precision depending on several factors, including the orbital regime, orbit coverage, quality and quantity of measurements, etc.

Specifying a very good predetermined accuracy for an orbit, in the context of this study, is thus always relative to the reference trajectory and by no means to the (non-observable) true path an object follows. The range of possible solutions for the catalogued reference trajectories was discussed in Section 1.2.

4.1.3. Assessing the accuracy relative to a reference

For a method thought of as providing a product of predetermined accuracy, it is important to understand, what the given accuracy figure is referring to. In principle, the closeness of agreement to the reference trajectory can be interpreted in many ways, so that it is essential to select an appropriate method to assess it.

For instance, one could compare the reference trajectory to the obtained orbit solution with a certain accuracy at a given point in time: the orbit determination epoch, after one day of propagation, after one revolution, etc. However, such an approach has the drawback that the selected instant in time might be associated with a significantly different solution when comparing with another time, e.g. after half a revolution - it can be even worse, considering the periodic nature of the osculating elements.

It should thus be the goal to define the method such that the predetermined accuracy is meaningful to a broad range of services working with the product. Observing that services like Collision Avoidance, Search and Initialisation, or Re-entry Prediction typically need ephemerides covering a span of several days in advance, it seems to be appropriate to derive the closeness of agreement between reference and product for a given time span.

Root Mean Square

A convenient way to assess the accuracy for a given time span is to compute the RMS of the component differences in an orbit-centered reference frame (radial, along-track and cross-track) for a pre-defined number of time steps n:

$$\Delta r_j^{RMS} = \sqrt{\sum_{i=1}^{n} \frac{1}{n} \left(r_{i,j} - \hat{r}_{i,j}\right)^2}, \; j \in \{U, V, W\}, \tag{4.1}$$

$$\Delta v_j^{RMS} = \sqrt{\sum_{i=1}^{n} \frac{1}{n} \left(v_{i,j} - \hat{v}_{i,j}\right)^2}, \; j \in \{U, V, W\}, \tag{4.2}$$

where $\hat{r}$ and $\hat{v}$ denote the radius and velocity vector of the reference trajectory, and the coordinates U, V and W refer to the radial, along-track and cross-track components, respectively. The predetermined accuracy can now be defined, for instance: "*The RMS of the difference between the provided trajectory and the catalogue reference in the along-track direction is in the range between 100 m to 500 m for a time span of 7 days.*"

While the RMS is defined component-wise, there can be difficulties in defining the threshold values independently, especially in view of correlations between the components. This problem still prevails, even if a combination like the Residuals Sum of Squares (RSS) is used, e.g.:

$$\Delta r^{RSS} = \sqrt{\sum_{i=1}^{3} \left(\Delta r_i^{RMS}\right)^2} \tag{4.3}$$

Encompassing volume

A method commonly used as a first step in the operational collision avoidance, is to define a perimeter around a target satellite and screen for chasing objects entering such an encompassing volume for a given time span (e.g. Anz-Meador (2004), Klinkrad et al. (2005)). Typically, one would define maximum values in an orbit-centered reference frame, i.e. in radial, along-track and cross-track direction. The famous *pizza box* used in the conjunction screening for the ISS is defined with 2 km up and down in radial direction and 25 km in along- and cross-track directions[1].

The degraded orbit solution would then be provided such that the difference to the reference trajectory will be always within the defined volume for the time span the trajectory is provided for.

Statistical distance

Being a measure of the distance between a given point and a statistical distribution, the *Mahalanobis distance* (Mahalanobis, 1936) is a metric which takes into account the uncertainties in the reference orbit. It is defined as:

$$d_M = \sqrt{\Delta \mathbf{x}^T \cdot \mathbf{P}^{-1} \cdot \Delta \mathbf{x}}, \tag{4.4}$$

here, in the context of this thesis, $\Delta \mathbf{x}$ would be the difference between the reference orbit and the solution with degraded accuracy. The covariance matrix $\mathbf{P}$ of the reference orbit is directly available from the object catalogue.

This will automatically result in a lower Mahalanobis distance for reference orbits that come with higher initial uncertainties from the orbit determination, while better solutions will also require a closer fit of the method with adapted accuracy for the same distance threshold.

4.1.4. Further method requirements

In Section 1.3.1, the rationale for the method to provide solutions of predetermined accuracy was outlined. Besides the aspects of commercialisation through the generation of diversified products with different accuracy from one single catalogue solution, orbit information might be of *dual-use*. This means that the highly accurate solution should not be disclosed to any other party without the appropriate clearance.

One can thus impose the requirement to design a method that does not allow to deduce the original (highly accurate) orbit solution from the provided (degraded) orbit information.

[1]These values were also used during the LISA Pathfinder Launch and Early Orbit Phase (LEOP) in the routine screening by ESA's Space Debris Office in December 2015.

Also, the method should be designed to be as flexible as possible in selecting the method metric (e.g. RMS ranges for the vector components or the Mahalanobis distance) as well as the fit span used in the comparison with the reference trajectory.

4.2. Two-line elements

The most prominent example of orbit information with reduced accuracy are the TLE provided by USSTRATCOM. Being based on the analytical SGP4/SDP4 theory, TLE made up the GP catalogue from the early 1970s (Vallado et al., 2006a) and ran in parallel with the SP catalogue from the late 1990s (Schumacher and Hoots, 2000). With significantly improved computational performance, the highly accurate SP vectors finally took over as the primary data source and there was no need anymore to run both catalogues in parallel.

From 2013 onwards, USSTRATCOM started to derive TLE from the orbits in the SP catalogue (Bowman, March 2014). Having to maintain only one catalogue was not the only advantage, however: with the original observations being already pre-processed and thus smoothed by a numerical theory, the analytical fit on the SP solution was shown to have a much better chance of following the object's orbit (Wilkins et al., 2000).

Although one cannot state that TLE are provided with predetermined accuracy, and indeed it is a very challenging task to assess that accuracy, it is a similar concept: There is a catalogue containing highly accurate information (which is not public and can only be accessed via dedicated Orbital Data Requests (ODRs) for specific events) with publicly available TLE data derived from that catalogue. A batch least squares technique is used to estimate the doubly averaged Kepler elements for the orbit update, similar to the method proposed in this thesis.

An example for deriving TLE parameters is shown in Figure 4.2 for ESA's ERS-2 mission. Using POE as the reference, a fit span of about five orbits[2] with data in 30-second-steps served as the input to update a TLE from the previous day (TLE epoch: 1995-09-06 17:45:07). In 1995, the POD for ERS-2 was based on SLR and altimeter data (the new Precise Range And Range-rate Equipment (PRARE) instrument augmented the products from 1996), with the residuals in the radial component being in the order of 5 cm (Ries et al., 1999).

For the estimation process, the weight matrix $\mathbf{W}$ was diagonal with all radius components weighted by 1.0 and all velocity components by 10^{-4}, similar to the approach in Flohrer et al. (2008).

It can be seen in Figure 4.2 that the residuals for the individual position components are in the expected order of magnitude when comparing with the figures from Table 1.1[3]: the RMS is 97 m, 237 m and 146 m in U, V and W direction, respectively.

It is also interesting to note that at the selected reference epoch (1995-09-07 01:00:00) there is already a deviation in all three components, which emphasizes that the Differential Correction (DC) is an optimisation for the whole fit span. Thus, even if the reference epoch would be set to the epoch of the SP epoch of the catalogued state, the DC process

[2]ERS-2 had a revolution period of approximately 100 min.
[3]with ERS-2 being at an altitude of $h \approx 780$ km, $i = 98.5°$, $e < 0.1$.

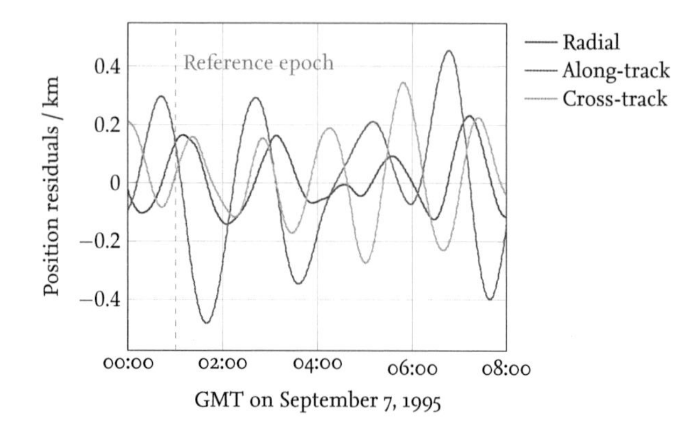

Figure 4.2.: Example showing the update of a TLE data set from 1995-09-06 17:45:07 to a reference epoch set to 1995-09-07 01:00:00. POE from ERS-2 in 30-second-steps were used as pseudo-observations for a fit span of about 7 hours and 40 minutes. The difference between the analytical SGP4/SDP4 theory and the POE is shown in the radial, along-track and cross-track components for the fit span.

will make sure that it will be adapted accordingly in favour of reaching an optimisation minimum for the entire fit span.

4.3. Orbit information with scaleable accuracy

The algorithm to derive orbits with predetermined (or scaleable) accuracy shall be based on modifying the geopotential degree and order of the propagator and then perform a least squares fit on the reference orbit with the modified force model. It is called GAMBIT and works as shown in Figure 4.3. It starts with selecting a fit span for a given reference orbit, where the latter can be based on a propagation of a catalogued state vector with full force model. In order to match a predetermined accuracy, e.g. given via an RMS (see Section 4.1.3), the geopotential degree and order are reduced and a DC fit is initiated. If the method converges in the envisaged accuracy range, the iteration is considered as successful and is stopped. Otherwise, it can be assumed that the selected values for the geopotential degree and order are not optimal, and the process is repeated, starting with a new modification of the force model by either increasing or decreasing the geopotential degree and order.

The method stops with no success, as soon as the maximum number of iteration steps has been reached. In addition, there could be a combination of a given reference orbit and a specified accuracy range that does not have a solution, which means that no suitable pair of geopotential order and degree is available (not shown in Figure 4.3). This also represents an unsuccessful stopping criterion.

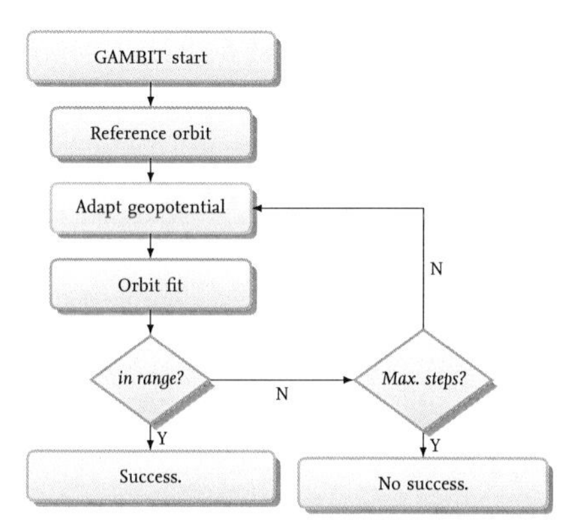

Figure 4.3.: Basic algorithm outline to derive an orbit with GAMBIT.

4.3.1. Modifying the geopotential

The first question to be adressed is related to which accuracy levels can be achieved. The geopotential is defined for integer values for the degree and order in the series expansion. Therefore, one cannot achieve any arbitrary difference, e.g. in RMS, between the reference trajectory and the desired orbit solution.

For the LEO region, an example for near-circular orbits with an inclination of 30° and varying altitude is shown in Figure 4.4. While for less accurate solutions, e.g. for a geopotential of 2×2 or 3×3 (degree $\times$ order), the step in accuracy degradation is quite significant, a much higher resolution is obtained for higher values of n.

It is possible to get a higher resolution by not having the restriction of a *symmetric* geopotential (i.e. $n = m$). This would mean to truncate the geopotential series for a given m_{max} with $m \leqslant m_{max} \leqslant n$. An example is shown in Figure 4.5. Again, circular orbits in the LEO region have been analysed, this time with an inclination of 54°. The symmetric potential solutions are shown with dashed lines in Figure 4.5. It can be seen that the regions in-between those solutions can, to some extent, be covered by omitting certain tesserals.

An example for the position residuals of such an orbit fit for a given RMS in the radial component of 30 m to 80 m is shown in Figure 4.6, computed for the orbit of Sentinel-1A[4]. It can be noted that the radial position residuals show variations of up to 136 m, while the along-track and normal component residuals can reach up to 517 m and 290 m, respectively. This was obtained for a 3×3 geopotential based on a reference trajectory with $n_r = 36$.

4.3.2. Analysis of method parameters

The methodology outlined in the previous section presupposes the proper selection of a set of algorithm parameters. First of all, convergence for a standard least squares ap-

[4] 693 km altitude, 98.2° inclination

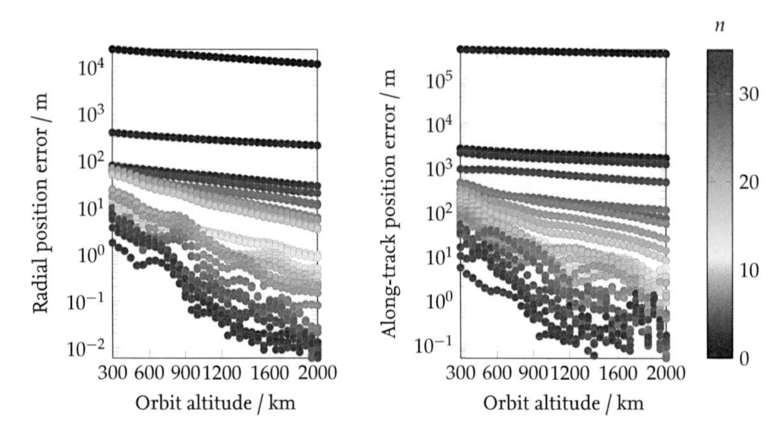

Figure 4.4.: Achievable accuracy levels for a symmetric geopotential ($n \times n$), comparing a reference trajectory (36×36) with solutions of varying geopotential degree and order for different altitudes. All orbits are near-circular and have an inclination of $i = 30°$.

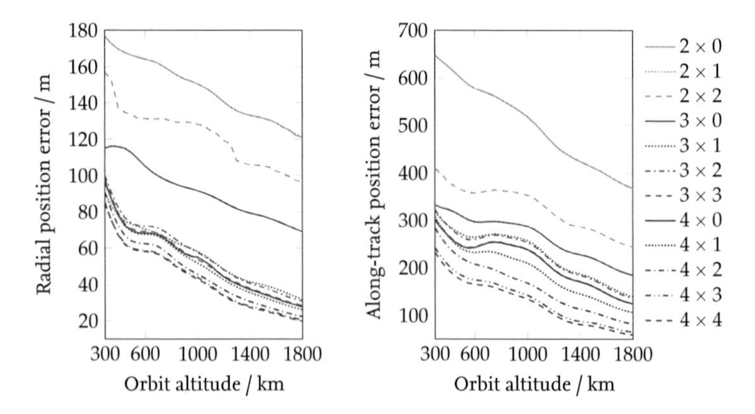

Figure 4.5.: Varying order of geopotential for fixed n. Here, $n \times m$ means that all tesserals are included for $n_i < n$, while only the contributions due to tesserals at degree n are omitted. The reference trajectory was propagated with $n_{ref} = 24$ for an inclination of 54°.

proach, as introduced with the differential correction method, is not always guaranteed to converge. One approach to improve this is the Levenberg-Marquardt method. Its influence in finding the locally optimum solution, shall be demonstrated in the following.

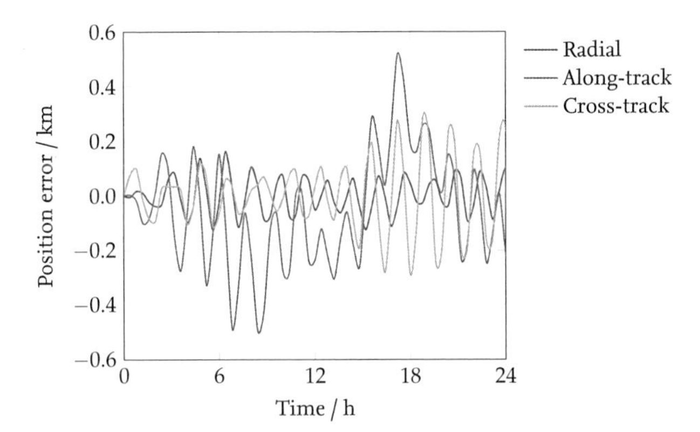

Figure 4.6.: Example of applying the algorithm to a real orbit. A TLE-derived state for Sentinel-1A was used to generate the reference trajectory. The success criterion was RMS-based for the radial component, restricting it to the range 30 m to 80 m. For a 3×3 geopotential the radial RMS was 68 m, along-track 204 m and cross-track 116 m. Reference trajectory: full force model, $n = 36$; Fit for 15 orbits, 307 samples (pseudo-obs).

Levenberg-Marquardt

The differential correction method is searching for the minimum of the cost function, as defined via Equation 3.32, by iteratively adapting the state vector. The search space has at least six dimensions for finding the optimum position and velocity vector. A visualisation by restricting it to only two of the position vector coordinates is shown in Figure 4.7 and serves as a strong argument in explaining the effectiveness of the LM in the orbital mechanics context. It can be seen, that the gradient in radial direction is extremely steep compared to the normal and along-track directions. This leads to a valley perpendicular to the radial direction, along which the minimum has to be found. One would thus prefer a method which takes larger steps along the valley and small steps in radial direction, in order to not step across the minimum. The search for the least squares solution via the Gauss-Newton method does exactly the opposite, which is the reason for convergence problems when using this method alone.

For the LEO region, an analysis was performed, first without and then with the LM method, to illustrate the effect of the LM damping parameter. The results are shown in Figure 4.8. It can be seen, that a pure differential correction (Gauss-Newton method) encounters problems for the low-inclination orbits, where convergence failed in all those cases where the z-values are negative in Figure 4.8.

Using a starting value of $\lambda_{LM} = 10^{-8}$ for the LM already solved most problems, as can be seen in Figure 4.8b. For the remaining cases, where a failed scenario is indicated, there

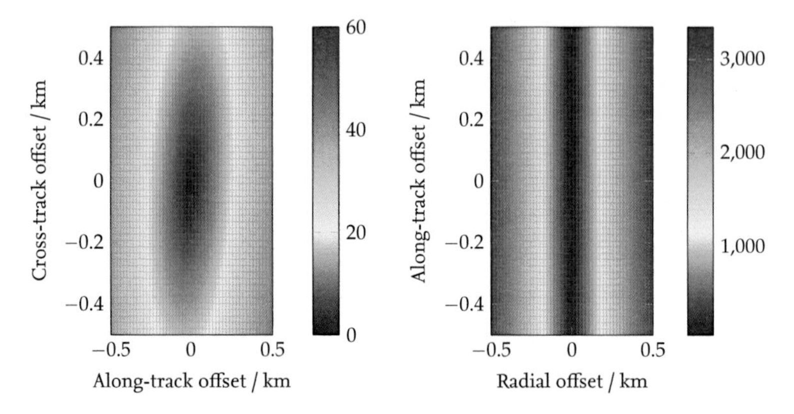

Figure 4.7.: Value of cost function (or performance index) as defined in Equation 3.32. A 12×12 geopotential fit was computed based on a reference trajectory with the same geopotential degree and order.

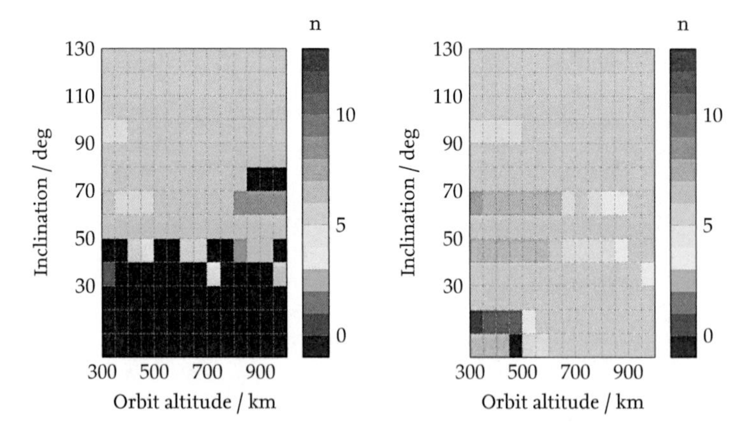

Figure 4.8.: Difference between computing the non-linear least squares solution with (b) and without LM (a) in the LEO region. The required geopotential degree was iteratively searched for based on an RMS criterion for the radial component, which had to be between 30 m to 80 m.

was no solution found between 30 m and 80 m for the RMS in the radial component. This only indicates, that for a symmetric geopotential, this interval was selected as too narrow.

A variation of the damping parameter was also analysed and it was found that already for $\lambda_{LM} = 10^{-6}$, the situation deteriorates, and many low-inclination bins cannot be solved, similar to the example without LM in Figure 4.8a.

Mahalanobis distance example

While in the previous sections, the RMS was used as the metric for finding the geopotential order and degree of the reduced force model, it is also possible to use a more statistically motivated criterion: the Mahalanobis distance, as introduced in Section 4.1.3, is then required to stay below a certain maximum value for the desired solution. This allows to have one single threshold, which takes into account the variances of the reference orbit and the associated correlations.

The Mahalanobis distance, in fact, can be geometrically interpreted as a way of transforming Euclidean distances of multivariate variables taking according to their variances. While for a given Euclidean distance in three-dimensional space all points from the center describe a sphere, the Mahalanobis distance would correspond to all points being on the surface of an ellipsoid. If the variances in all three directions are unity, the Mahalanobis distance is equal to the Euclidean distance.

Using an ephemeris and an associated covariance for ESA's Cryosat-2 satellite[5], the reference was generated using a full force model with a 36×36 geopotential. The Mahalanobis distance was then computed (via Equation 4.4) as the statistical distance of an orbit with reduced force model with respect to the propagated covariance matrix of Cryosat-2. The results are shown in Figure 4.9a. While for a 12×12 geopotential, the solution stays still very close to the reference, one can observe a drift by further reducing the force model, which is due to the neglected zonal harmonics.

An advantage of such an approach is that it directly scales with the initial uncertainty in the reference orbit solution: If the orbit determination result for the reference comes with high covariance, a given threshold for the Mahalanobis distance will assure that such an orbit will not be degraded as much as would be the case for an orbit with high accuracy.

In Figure 4.9b, the evolution of the uncertainties in the components of the radius vector shows, that the scaling is dependent on the propagation: especially the along-track error grows secularly with time. This effectively means, that for states being close to the epoch (and thus close to the initial covariance), the Mahalanobis distance is computed for a reference trajectory with less uncertainty, as opposed to states far from the epoch.

Fit span and sample size

From the orbit determination two more parameters are known that may have a strong influence on the solution: the fit span and the sample size, or the number of pseudo-observations within that span. In order to determine how different values affect the solution, a sensitivity analysis was performed. The results are exemplarily shown for the Cryosat-2 orbit (as given in the previous section) in Figure 4.10. While very short fit spans in the order of one orbital revolution are very sensitive to variations in the fit span, this

[5]Mean altitude: 717 km, Inclination: 92°.

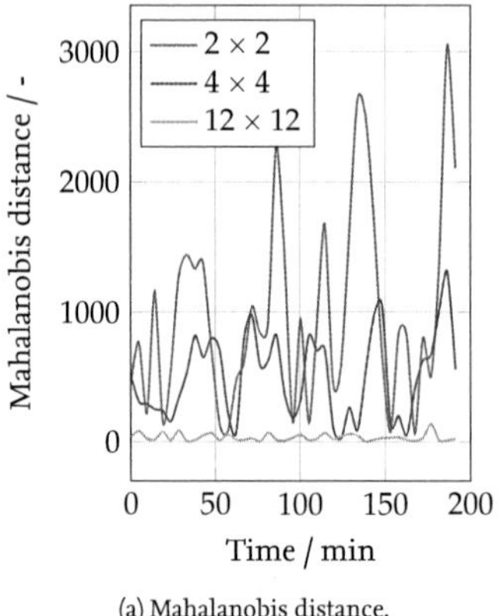

(a) Mahalanobis distance.

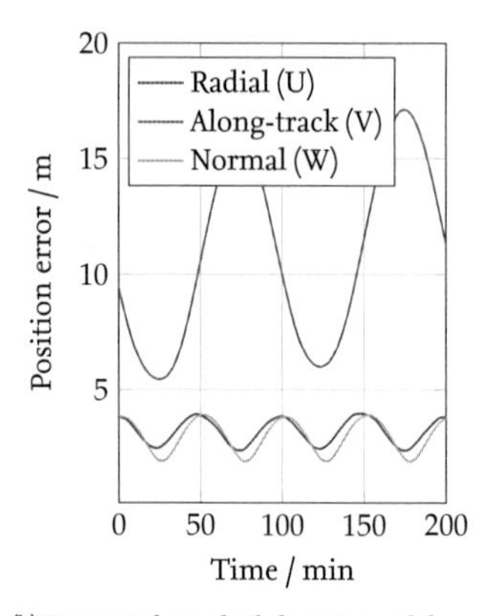

(b) Propagated standard deviations of the radius vector.

Figure 4.9.: Example for the evolution of the Mahalanobis distance for a geopotential of varying degree and order. The reference trajectory had a geopotential of 36 × 36. The propagated standard deviations of the radius vector, using an initial covariance matrix from a CDM, are shown in the right figure.

changes for fit spans across several orbits. It is very interesting to note that the selected number of pseudo-observations for a given fit span, on the other hand, does not strongly affect variations. This allows to draw the conclusion that there is a relatively large space from which the pair fit span and sample size can be selected without significantly affecting the performance of the algorithm.

4.3.3. Obtaining the covariance matrix

A great advantage of applying the DC method, which is also used in the orbit determination, is the fact, that the covariance matrix of the solution is readily available as the inverse of the Fisher information (Equation 3.37):

$$\mathbf{P}_{DC} = \left(\mathbf{H}^T \cdot \mathbf{W} \cdot \mathbf{H}\right)^{-1}. \tag{4.5}$$

However, this does not mean, that this is a real covariance matrix. First of all, it can only be a covariance matrix, if the weight matrix contains the reciprocal measurement errors in the observations (Tapley et al., 2004).

As in the presented method cartesian state vectors are used as pseudo-observations, one possible solution is to introduce another iteration after the first one has successfully converged. The second iteration would use the obtained RMS with respect to the reference trajectory as input to setup the weight matrix and perform the fit again.

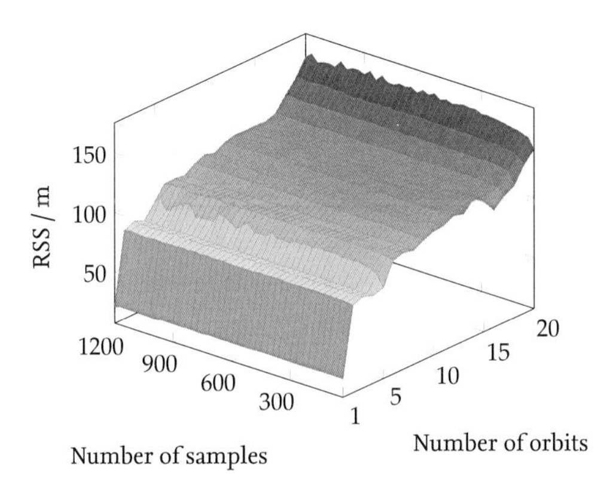

Figure 4.10.: Cryosat example: showing how fit span and number of sampled points affect the solution.

Having a weight matrix that directly reflects the (pseudo-)observation residuals via the RMS of the fit is only the first step: it is also important to combine this estimate with the original covariance in the reference orbit. There are a few difficulties in doing this. First of all, it is important to assess the correlation between the original covariance matrix and the one obtained from the fit. By reducing the geopotential degree and order for a subsequent fit, the uncertainties in the reference orbit are not taken into account. The least-squares state updates require the state transition matrix, which, in principle, is the same as in the propagation of the covariance of the reference orbit across the fit span. It thus reflects the orbital mechanics only, being similar for both orbits. Correlation in single components can occur: an example is the analysis by Matney et al. (2004) for the along-track error in drag-affected orbits.

Analogously, one can imagine orbits, where errors from the truncation of the geopotential series or even the residuals in the harmonic coefficients are pronounced in the reference solution. When performing a fit, it can happen, that this effect is amplified by cutting at even lower degree and order. However, as the covariance of the fit solution is not propagated, this effect can be considered as small.

Assuming, that correlations between the elements of the covariance matrices can be neglected, the combination of the two can occur according to the scheme shown in Figure 4.11. The covariance of the fit is constant throughout the fit span, as it reflects the residuals of the fit with respect to the reference. The RMS is not affected by the selection of the epoch within that span.

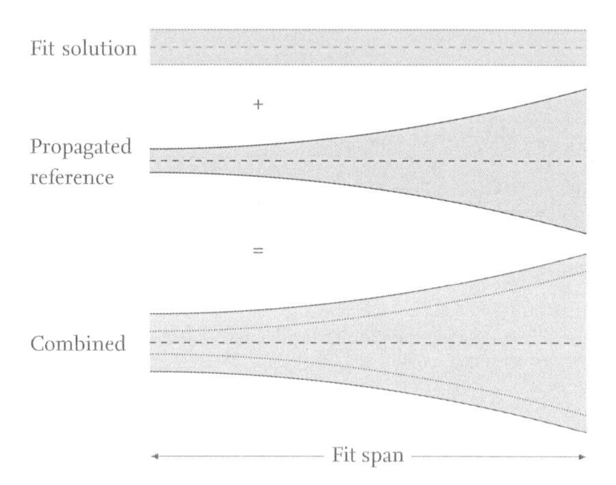

Figure 4.11.: Scheme showing how the covariance matrix of the reference orbit and the one obtained from the least squares solution can be combined.

One can thus add the constant covariance of the fit to the propagated reference covariance, which degrades with time, if the fit is based on a propagated reference trajectory, to obtain the combined covariance matrix $\mathbf{P}_{cmb}$:

$$\mathbf{P}_{cmb}(t_i) = \mathbf{P}_r(t_i) + \mathbf{P}_{DC}(t_0) . \tag{4.6}$$

4.3.4. Applying the method to different orbit regions

So far, only orbits in the LEO region have been discussed. For a general applicability, it is also important to study the implications for other important orbital regions.

GEO

The magnitude of the accelerations due to the dominant central body term of the geopotential declines with $1/r^2$. Thus, the effectiveness of the GAMBIT method can be considered lower at geostationary altitude. In Figure 4.12 the along-track and cross-track RMS for two different fits as a function of the geographic longitude are shown. For a 2×2 fit, the along-track error can reach up to several 100 m, while already for a 3×3 fit, the errors with respect to a 36×36 trajectory are reduced to the 10 m level. The cross-track error in Figure 4.12b is on the 1 m level for a 2×2 geopotential and also decreases by an order of magnitude for a 3×3 fit.

Considering, that 170 m in GEO correspond to about 1″ for ground-based optical observations, this is already in the order of the measurement noise for typical sensors (e.g. Kelecy et al. (2008)).

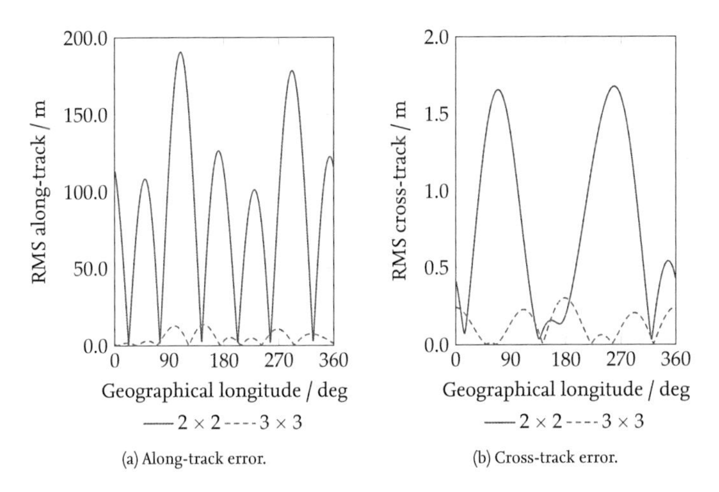

(a) Along-track error.

(b) Cross-track error.

Figure 4.12.: Fitting a reference orbit in GEO ($n = 36$) with a reduced geopotential for varying geographical longitude. A fit span of four orbits and 83 samples were used.

MEO

The next analysis was performed for the MEO region. In Figure 4.13 a 2×2 and a 3×3 fit are shown as a function of altitude, encompassing all current navigation constellations. It is very interesting to note, that the radial and, even more pronounced, the along-track error both show maxima at the operational altitudes of the different constellations. This can be explained by the geosynchronous nature of those orbits, where higher-degree terms in the geopotential are required to consider the resonance cases. For example, the GPS constellation has a repeating groundtrack every two orbits, while satellites in the Globalnaya navigatsionnaya sputnikovaya sistema (GLONASS) constellation fly over the same regions every eight orbits. For Beidou, it is seven orbits, while satellites in the Galileo constellation repeat their groundtrack every ten orbits.

Eccentric orbits

The properties of the method when applied to orbits with high eccentricity were also studied. For varying the perigee altitude and the eccentricity, the required degree for a symmetric geopotential, based on an RMS for the radial component between 50 m to 200 m, is shown in Figure 4.14a. The first remarkable result is that for a wide range, up to eccentricities of $e = 0.6$, the solution was always a 2×2 geopotential fit on the reference. However, when the eccentricity is further increased, up to GEO altitudes ($e \in (0.706, 0.730)$ in the domain of the shown perigee altitudes) and even above, the required geopotential degree steeply increases.

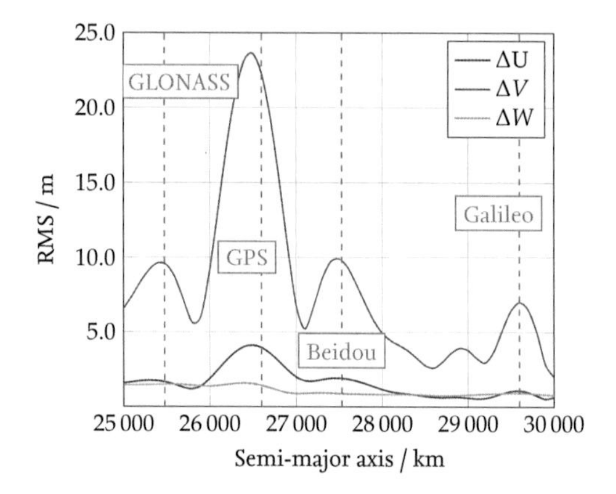

Figure 4.13.: Fitting a reference orbit in MEO ($n = 36$) with a reduced geopotential (3×3) for varying altitude, covering the operational orbits of navigation satellite constellations. The fit span was eight orbits with 163 samples in total.

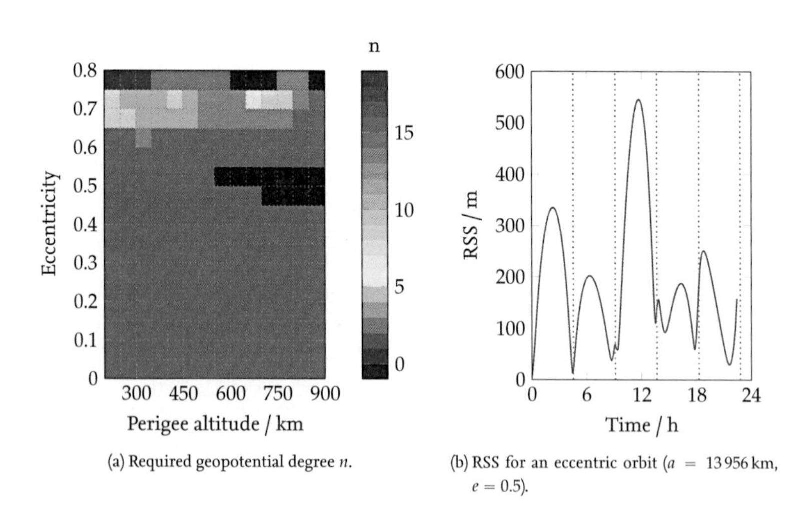

(a) Required geopotential degree n.

(b) RSS for an eccentric orbit ($a = 13\,956$ km, $e = 0.5$).

Figure 4.14.: Required degree for a symmetric geopotential (left) in order to obtain a fit with an RMS of the radial position component between 50 m to 200 m. A full force model ($n = 36$) was used as a reference. The fit span was about 5 orbits with about 24 samples per orbit. Apogee passes are shown with blue dotted lines (right).

One can also see in Figure 4.14a, that there were combinations for perigee altitude and eccentricity, where the iteration did not converge to a solution. For the field with $0.45 < e < 0.55$ and $550\,\text{km} < h_p < 900\,\text{km}$, it failed because even for a geopotential of 2×2, no solution could be found - the RMS in the radial position component was less than 50 m. Further reducing would mean to use a two-body gravity field. This, however, results in significantly higher deviations, even above the defined upper limit.

Finally, some simulations with $e = 0.75$ also failed. The iteration went until similar values for n were reached as for the adjacent altitude bins. With the solution still being above the defined upper limit for the radial component, the next increment in n led to a jump in the RMS below the defined lower limit. Thus, there is no symmetric geopotential, which can provide a solution under the defined conditions.

The problem for high-eccentricity orbits with $e > 0.6$ is the distribution of sample points along the orbit. For the analyses shown here and in the previous examples, the steps were equidistant in time. With increasing eccentricity, the orbit region around the apogee gets a stronger weight with respect to the perigee region, simply because there are more samples around the apogee. This is underlined by the RSS in the position vector, which is shown exemplarily for one of the failed orbits in Figure 4.14b. The RSS is very low near the apogee passes (indicated by a dotted line), while it reaches a maximum near the perigee passes.

A possible solution for eccentric orbits would be to use regularization, an approach well known for eccentric orbits, for example the generalized Sundman time transformation (Sundman, 1913). A possible implementation would then take equidistant steps in the mean anomaly.

4.3.5. Degradation by adding noise

An alternative and easy way of degrading orbit information is to introduce pseudo-random noise. Known as Selective Availability, such a procedure was used for the Navigation System using Timing and Ranging (NAVSTAR) constellation of GPS until 2001, where dithering applied to the GPS clocks resulted in position errors on the order of 100 m (Kremer et al., 1990).

Therefore, a method that adds pseudo-random noise to the individual components of the state vector was studied. An exemplary result is shown in Figure 4.15 for a circular 400 km orbit at 54° inclination. Gaussian noise with zero mean and a standard deviation of 100 m was added to the position components. This resulted in the scattered values for the osculating semi-major axis with respect to the *true* solution, as shown in Figure 4.15.

A third plot, however, shows that applying a smoothing allows for obtaining a significantly improved estimate of the semi-major axis. In this case, a simple centered moving average filter with 31 points was used. A similar procedure can be found in the literature for improving GPS signals by averaging.

Finally, a procedure that adds noise, will also add higher frequencies to the signal, an effect which is not desired in view of the subsequent interpolation envisaged for the so-

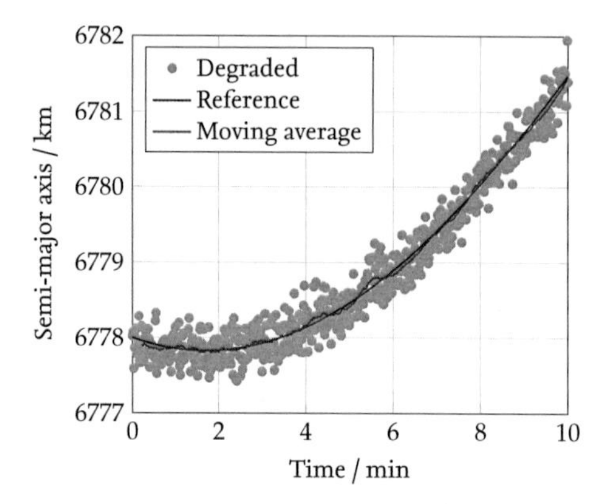

Figure 4.15.: Evolution of the osculating semi-major axis, for a circular orbit at 400 km altitude and 54° inclination. Gaussian noise with zero mean and a standard deviation of 100 m was added. Applying a moving average filter with 31 centered points allowed to obtain estimates close to the original values.

lution (cf. Figure 1.1). In principle, it should be beneficial to remove higher frequencies before going into the interpolation, in order to mitigate effects due to under-sampling.

5

Ephemeris compression to provide continuous data

The way orbit information is exchanged between different entities today can be principally attributed to one of the following two categories: Either information is provided for one single epoch, for example by means of a state vector and covariance, or a table of ephemerides is provided covering a certain time period. While the former method requires the user to perform his own extrapolation to any other point in time, the latter approach allows to interpolate between the given data points.

Both options have their advantages but also disadvantages: for example, TLE have to be processed by SGP4/SDP4, even if the state is to be obtained at the epoch. The software needs to be the same on both sides - for the generation of the TLE and their subsequent processing in applications. If the employed force models do not match, additional errors will be introduced.

Interpolation of tabulated data has the advantage that no orbit extrapolation (or propagation) software is required. Standard interpolation techniques can be used. However, the provided data comes with a certain time step and the user does not necessarily know, which frequencies are present in the underlying data to properly sample it with his interpolation. For data messages like the OEM (CCSDS 505.0-B-1, 2010), there are dedicated keywords that the distributing entity can use to provide information on which interpolation method and interpolation degree to use.

The method proposed here is to go one step further and let the interpolation already occur *before* the information gets distributed. Thereby, full control to minimise the additional error introduced by the interpolation is retained. With polynomial coefficients being provided, users do not need to care about the proper handling of the information, as the reconstruction of a polynomial series is a trivial task.

Several authors already studied applying polynomial interpolation to Earth orbits (e.g. for the GPS constellation (Horemuž and Andersson, 2006; Schenewerk, 2003), as well as more general analyses, e.g. (Deprit et al., 1979; Segerman and Coffey, 1998)) or even to interplanetary orbits, like the orbit solutions provided as Chebyshev polynomials for the bodies of the solar system by JPL (Seidelmann, 2006).

The latter approach was identified to be very promising for the application to any Earth orbit (Deprit et al., 1979). Therefore, the next section will give a brief overview on the properties of Chebyshev polynomials. Then, several examples for the interpolation of radius and velocity vectors for different orbits will be shown.

Providing orbit information as polynomial coefficients has the additional advantage, that a considerable data compression can be achieved when compared with tabulated data. Moreover, an approach to compute an envelope function for the variances of the state vector components and interpolate it subsequently was studied and will be discussed.

Finally, polynomial interpolation is a well-suited post-processing step for orbits generated via the GAMBIT method, as those typically do not contain high frequency components of the geopotential, which makes interpolation easier. Furthermore, as the force model is adapted for each orbit, providing polynomials is less error-prone on the user's side. It shall be shown, that the polynomials can be used to approximate the input orbit information without introducing significant errors.

5.1. Chebyshev polynomials

The polynomial interpolation can be based on Chebyshev polynomials of the first kind, $T_k(t)$. The polynomial P_n interpolates the to be approximated function at $n+1$ nodes:

$$P_n(t) = \sum_{k=0}^{n} c_k \cdot T_k(t), \tag{5.1}$$

where c_k are the polynomial coefficients that are specific for any given function P. The Chebyshev polynomials of the first kind satisfy the following equation for degree n and argument $t \in [-1, 1]$:

$$T_n(t) = \cos(n \cdot \arccos t). \tag{5.2}$$

In order to compute the polynomials of higher order, a recurrence relation can be used (Abramowitz and Stegun, 1964):

$$T_{n+1}(t) = 2 \cdot t \cdot T_n(t) - T_{n-1}(t). \tag{5.3}$$

5.1.1. Chebyshev nodes

In the application, the interpolation nodes are represented by the ephemerides, the latter typically provided equidistantly with respect to time. However, these nodes do not coincide with the Chebyshev nodes, which reduce Runge's phenomenon[1] through the denser spacing of nodes near the interval borders. The Chebyshev nodes are the roots of the Chebyshev polynomials, thus resulting in:

$$t_k = \cos\left(\left(k + \frac{1}{2}\right) \cdot \frac{\pi}{n+1}\right), \; k = 0, \ldots, n. \tag{5.4}$$

[1] Polynomials of high degree show oscillations and high interpolation errors at interval borders for equidistant interpolation.(Runge, 1901)

As the Chebyshev polynomials are defined for $-1 \leqslant t \leqslant 1$ only, the independent variable t^* has to be converted first, using the following equation with the lower and upper interval border of the ephemerides table given as a and b, respectively:

$$t = \frac{2 \cdot t^* - (a+b)}{b-a}. \tag{5.5}$$

5.1.2. Chebyshev polynomial coefficients

The orthogonality property can be used to compute the coefficients of the polynomial given in Equation 5.1, given the fact that this polynomial is equal to the ephemeris (function) value at the Chebyshev nodes, $P_n(t_k) = f(t_k)$. Therefore, $f(t_k)$ is multiplied with a Chebyshev polynomial of the first kind and summed over the $n+1$ nodes (Gil et al., 2007):

$$\begin{aligned}\sum_{k=0}^{n} f(t_k) \cdot T_l(t_k) &= \sum_{i=0}^{n} c_i \sum_{k=0}^{n} T_i(t_k) \cdot T_l(t_k) \\ &= \frac{n+1}{2} \cdot c_l,\ l > 0\end{aligned} \tag{5.6}$$

Thus, for $l > 0$, the coefficients c_l can be computed as

$$c_l = \frac{2}{n+1} \cdot \sum_{k=0}^{n} f(t_k) \cdot T_l(t_k),\ l > 0, \tag{5.7}$$

and for $l = 0$:

$$c_0 = \frac{1}{n+1} \cdot \sum_{k=0}^{n} f(t_k) \cdot T_l(t_k). \tag{5.8}$$

5.2. Ephemeris compression

5.2.1. Reference orbits

The performance of interpolation and compression methods has to be assessed with respect to some reference. A convenient approach is to generate a highly accurate reference trajectory through propagation: NEPTUNE was used with initial conditions obtained from several different TLE to obtain reference orbits for different orbital regions.

The data shown in Table 5.1, was used in the following for single satellite examples.

The propagator was configured to use a full force model (in this case a 24×24 geopotential, atmospheric drag (NRLMSISE-00), lunisolar perturbations, solar radiation pressure in combination with a conical umbra / penumbra shadow model, solid Earth and ocean tides), the propagation of the covariance matrix included J_2 contributions to the variational equations. As TLE do not come with any uncertainty information, the exemplary full covariance matrix (diagonal and off-diagonal elements) was taken from an operational case used by ESA's Space Debris Office collision avoidance service. Each orbit was propagated for at least 14 revolutions.

Note that although the SGP4 force model does not match the one of the numerical integration, using the osculating TLE states as initial state vectors for the propagation was considered as viable for providing representative results.

Table 5.1.: Satellites selected for the examples in this chapter, with doubly averaged perigee altitude h_p, eccentricity e and inclination i from TLE. Osculating states were derived from TLE and directly used as initial states in the numerical propagation.

Name	Short	h_p / km	e	i / deg
ATV-2	ATV	359.0	0.0019	51.6
Sentinel-1A	S1A	695.0	0.000 14	98.2
Galileo-8	GL8	23 214.0	0.000 27	55.1
Ariane-5 R/B	AR5	252.0	0.7282	5.9
Meteosat-10	M10	35 788.0	0.000 02	0.06

5.2.2. Results

It is essential to properly select the interpolation interval length and the polynomial degree in order to approximate the reference trajectory with a given accuracy. The frequencies present in the reference trajectory have to be known in order to select the proper sampling. An additional source of error is referred to as the *Gibbs phenomenon* (Gibbs, 1898): signals containing discontinuities can only be approximated to a certain extent by a series of continuous functions. However, if the polynomial interpolation is applied to an already smoothed signal, which is the case here for the numerical reference orbit or the orbit with predetermined accuracy, there will not be any discontinuities.

For the selection of interval length and polynomial degree, Seidelmann (2006) gives an example: the JPL ephemerides of the Earth are segmented into 16 d intervals with a polynomial degree of 12. Depending on which DE model is used, the interpolation error for all coordinate values might be less than 0.5 mm (Seidelmann, 2006). This accuracy should not be confused with the orbit determination residuals - the interpolation error is always relative to the reference trajectory.

In presenting the transition from a GP to an SP catalog, (Schumacher and Hoots, 2000) states that an SST system may provide compressed ephemerides with accepted interpolation errors for the position vector of up to 100 m.

Some analysis results for three different Accepted Error Levels (AELs) are presented in the following: 1 m, 10 m, and 100 m. An AEL of 1 m is the range RMS with respect to a reference trajectory. From an SST perspective, such a product would be appropriate even for the collision avoidance service. Providing interpolated SP vectors with residuals up to 100 m, on the other hand, are sufficient for ground-based tracking purposes.

The residuals presented in the following were computed for the position (geometrical range) at one minute steps in the GCRF. The first result is shown in Figure 5.1 for the LEO region. The required polynomial degree is given as a function of granule length and orbit altitude (near-circular orbits assumed) for an AEL of 1 m. This result confirms the applicability of Chebyshev interpolation to perturbed orbits - even for very low altitude

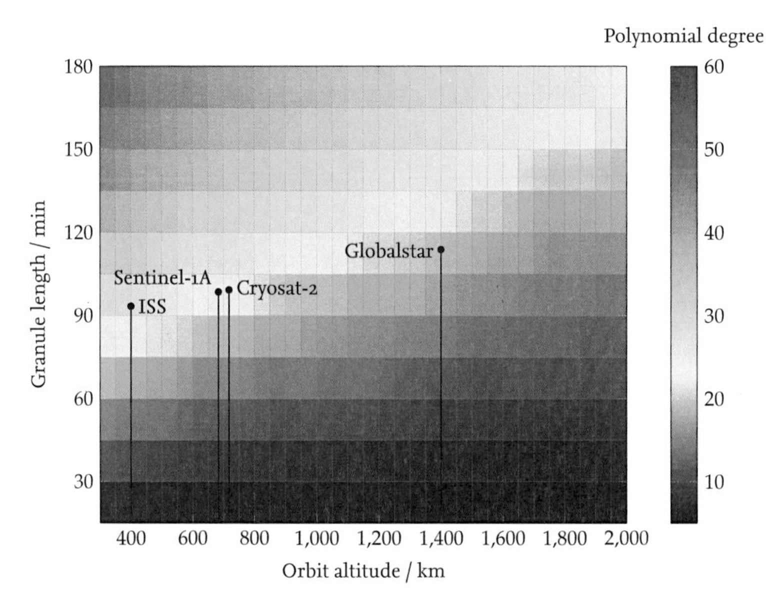

Figure 5.1.: Polynomial degree as a function of orbit altitude and granule length in the LEO region. An inclination of 54° was used, which is close to two of the exemplary missions shown, ISS and Globalstar. Two additional ESA missions (with higher inclinations!) are provided for comparison. A full force model was used. The blue dots mark the orbit altitude and period.

orbits with significant drag contributions. For a granule length close to the orbital period (e.g. about 90 min for the ISS) the required polynomial degree is about 20.

In Figure 5.2 the relationship between polynomial degree and granule length is shown for the different AEL and the reference objects defined in Table 5.1. It is quite interesting to note that there is a linear relationship between both quantities. This allows for a simple approximation of the number of required polynomial coefficients to cover a certain interpolation interval, as shown in Table 5.2, irrespective of the segmentation applied to the entire propagation span. Note that for a complete orbit, with the state vector containing six elements, the figures given in Table 5.2 have to be multiplied by six. For example, the orbit of Sentinel-1A (with an AEL of 100 m) would require 34.9 coefficients per hour, or 5860 coefficients to cover a whole week.

While the results shown above were determined for a maximum accepted error in position, as specified by the AEL, the residuals in the single components, of course, may be less than this threshold, even up to several orders of magnitude. A selected example for Sentinel-1A with a granule length of 300 min and an AEL of 10 m is shown in Figure 5.3.

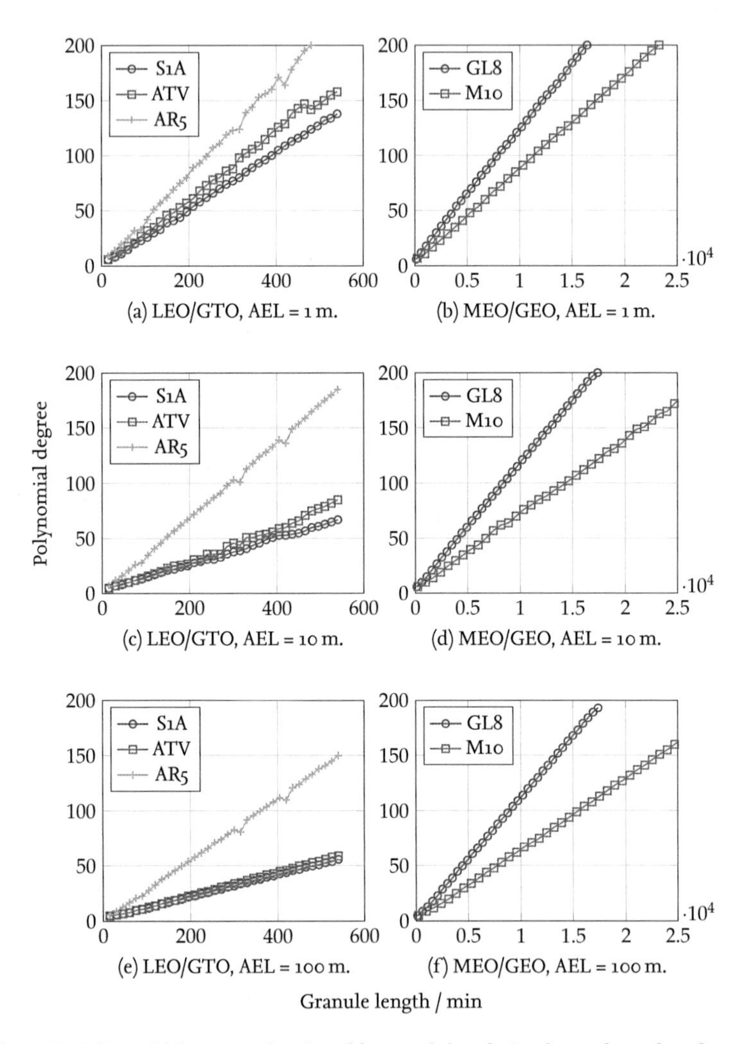

Figure 5.2.: Polynomial degree as a function of the granule length. Results are shown for reference orbits as given in Table 5.1, and three different accepted error levels for the position. Note that the x-axis for the MEO/GEO plots are scaled by 10^4.

In this example, one segment contains three orbital revolutions of Sentinel-1A. It can be

Table 5.2.: Number of required coefficients per hour of interpolation span, based on the results shown in Figure 5.2.

	Coefficients per hour		
	AEL = 1 m	10 m	100 m
ATV-2	17.4	9.12	6.18
Sentinel-1A	15.1	7.08	5.82
Ariane-5 R/B	24.6	20.3	16.5
Galileo-8	0.713	0.674	0.653
Meteosat-10	0.503	0.402	0.380

seen that the interpolation errors are higher for mid-interval values and smaller at the interval edges, as a result of a denser spacing of Chebyshev interpolation nodes.

Another interesting example is a high-eccentricity orbit as shown for the Ariane-5 upper stage in Figure 5.4. Here, the errors near the perigee are clearly dominating. It can also be seen that the residuals in single segments may be orders of magnitude below the accepted error threshold for the entire interval, while one single perigee pass may be pivotal in the determination of the required polynomial degree for the GTO.

So far, the interpolation was based on radius and velocity vectors in the GCRF. Another idea is to interpolate osculating classical orbit elements. Again, the same orbits were analyzed with the error threshold being the position. The only difference was that the Chebyshev polynomial coefficients were determined for the semi-major axis, eccentricity, inclination, right ascension of the ascending node, argument of perigee, and mean anomaly. However, it turned out that switching to Keplerian elements does not yield any advantage. For example, the interpolation of the orbit of Sentinel-1A was performed with polynomials of up to double the polynomial degree compared to the results in Figure 5.2 for the same granule length. A reason for this is that the interpolation has to be done for functions with higher frequency components having significant amplitudes. An example is shown in Figure 5.5 for the eccentricity interpolation of Sentinel-1A. Two examples for polynomials with $n = 20$ and $n = 40$ are compared to the reference trajectory. In this scenario, the required polynomial degree was $n = 78$ for a granule length of 300 min and an AEL of 10 m.

5.3. Covariance matrix compression

The covariance matrix provided by JSpOC in a CDM is valid only at the Time of Closest Approach (TCA). Analyses involving, for example, an avoidance manoeuvre for one of the two objects, would require a propagation of the uncertainties. One can use a similar argument as for the state vector information and ask for interpolated covariances. Alfano et al. (2004) analysed a method based on interpolating data points with quintic splines. While such an approach does not guarantee positive definiteness, Woodburn and Tany-

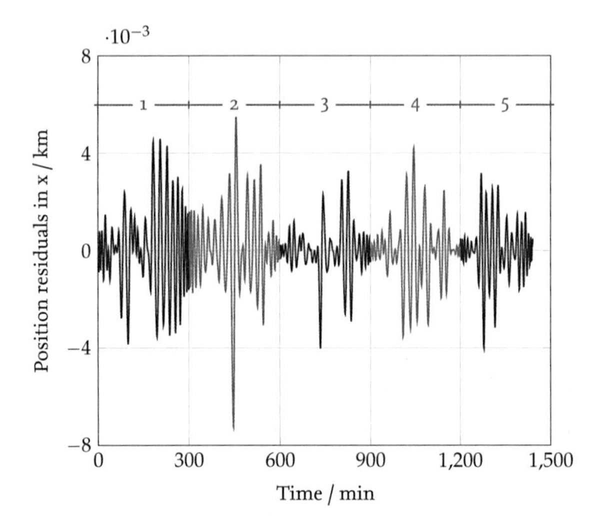

Figure 5.3.: Example for residuals in the x-component of the radius vector (GCRF) of Sentinel-1A, with a granule length of 300 minutes and an AEL of 10 m. Individual segments are labeled in red.

gin (2002) described a method based on the eigen-decomposition of a 3×3 matrix and subsequent interpolation of the sigma values and the eigenvector matrices. Tanygin (2014) gives another method, which can be directly used on the covariance matrix and guarantees positive definiteness.

In the context of this thesis, an approach was studied, where the interpolation of the variances (or diagonal elements of the covariance matrix) was performed on the envelope (supremum function) computed for those elements. The motivation was to significantly reduce the required amount of data to be transferred, while still preserving sufficient information for the intended purpose.

In Figure 5.6 an example is shown for the propagated variances of the position vector components in the object-centered reference frame, as well as an envelope for the along-track component. The main advantage of applying an interpolation algorithm on an envelope is that the latter does not contain oscillations and, by definition, is the supremum of the uncertainties in the individual directions. This will always result in a conservative estimate.

The definition of an envelope is only possible for the diagonal elements of the covariance matrix, i.e. the position and velocity variances. While some correlation coefficients in the off-diagonal elements have similar properties, others show oscillations, which means that the correlation (positive or negative) between different components is a function of the position along the orbit. An example is shown in Figure 5.7 for the correlations be-

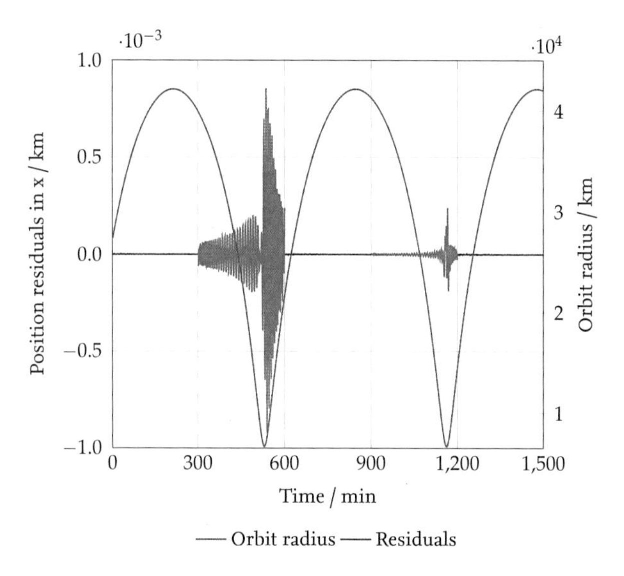

Figure 5.4.: Example for residuals in the x-component of the radius vector (GCRF) for the Ariane-5 upper stage in a GTO, with a granule length of 300 minutes and an AEL of 10 m.

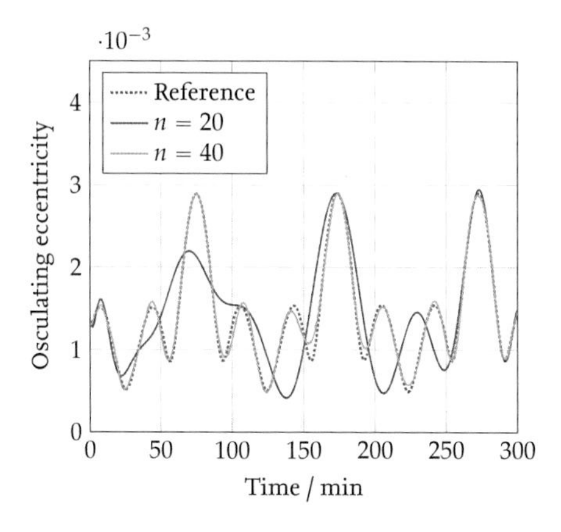

Figure 5.5.: Example for Sentinel-1A eccentricity interpolation. The reference trajectory is compared to polynomials of degree $n = 20$ and $n = 40$. The granule length was 300 min. For an AEL of 10 m, the required polynomial degree was $n = 78$.

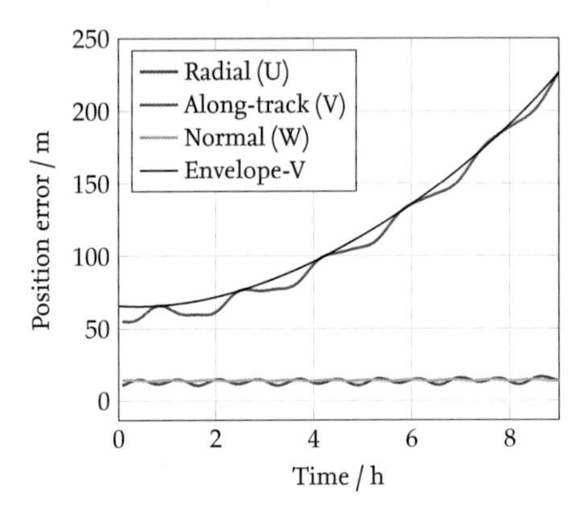

Figure 5.6.: Example for the propagation of a full covariance matrix, showing how the position variances in an object-centered frame evolve. A sun-synchronous LEO was used here, with a typical initial covariance matrix.

tween radial and along-track position, along-track and cross-track position, as well as the correlation of the cross-track position component with the radial and along-track velocity components.

Following the example in Figure 5.6, it can be seen that the envelope for both, the radial (U) and the normal (W) component are trivial and result in a constant supremum. Thus, the algorithm described hereafter will be shown for the transversal component but is likewise valid for the other directions.

Three different filters will be applied to the propagated variance, which essentially is a discrete time series to be analyzed for the determination of the envelope function $\mathrm{Env}(t)$. It is defined as the supremum of the variance in the along-track direction $V(t)$:

$$\mathrm{Env}(t) = \sup\left[V(t_i)\right],\ t_i \in [t_0, t] \tag{5.9}$$

Determine extrema and keep maxima The first step is to determine the extrema of the discrete time series, keeping in mind that the ultimate goal is to determine a set of points that can, in the end, be used for the interpolation of the envelope. The extrema are determined by evaluating finite differences:

- A local *minimum* is assumed at t_i, if:

$$V(t_{i+1}) - V(t_i) > 0 \wedge V(t_i) - V(t_{i-1}) < 0$$

- A local *maximum* is assumed at t_i, if:

$$V(t_{i+1}) - V(t_i) < 0 \wedge V(t_i) - V(t_{i-1}) > 0$$

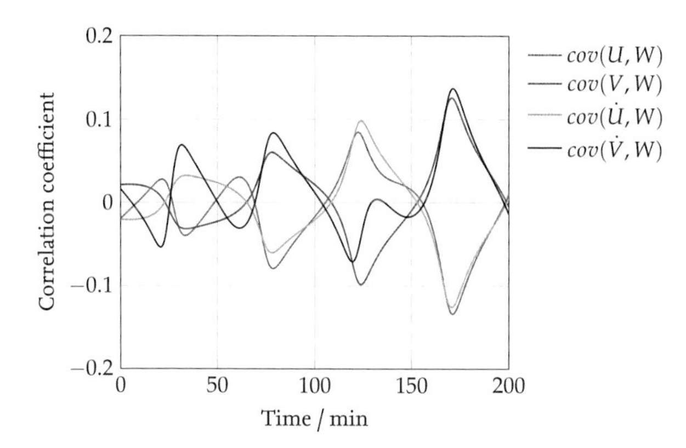

Figure 5.7.: Evolution of several correlation coefficients, showing clearly, that an envelope computation is useless, as soon as the oscillations come with sign changes.

- An *inflection point* is assumed at t_i, if:

$$\begin{aligned} &\{V(t_i) - 2V(t_{i-1}) + V(t_{i-2}) < 0 \wedge \\ &\ V(t_{i+2}) - 2V(t_{i+1}) + V(t_i) > 0\} \vee \\ &\{V(t_i) - 2V(t_{i-1}) + V(t_{i-2}) > 0 \wedge \\ &\ V(t_{i+2}) - 2V(t_{i+1}) + V(t_i) < 0\} \end{aligned}$$

The simple relations using finite differences are quickly identifying the extrema. The above formulation for the inflection point is an approximation of the second derivative of the function at t_i using second order backward and forward differences, respectively. For example, the backward differences $\nabla^2 V(t)$ are obtained via:

$$\begin{aligned} \nabla^2 V(t) &= [V(t_i) - V(t_{i-1})] - [V(t_{i-1}) - V(t_{i-2})] \\ &= V(t_i) - 2V(t_{i-1}) + V(t_{i-2}) \end{aligned}$$

with the result for the along-track component shown in Figure 5.8.

It is clear that not all extrema will be relevant for the interpolation of the envelope function. Therefore, the next filter step removes all identified local minima from the set of identified points. An additional step was to discard those inflection points which were detected following a local maximum. Finally, the set of remaining points having passed this filter step are shown in Figure 5.9.

Point shift filter With the extrema being identified and filtered for the relevant ones, the next filter will perform slight adjustments by shifting the remaining points, if a point is

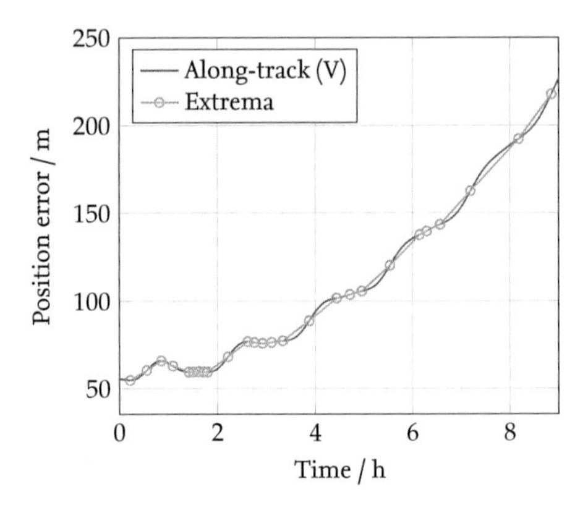

Figure 5.8.: Along-track error with extrema being identified. Linear interpolation for adjacent extrema, shown with orange line, will be used in subsequent filter steps.

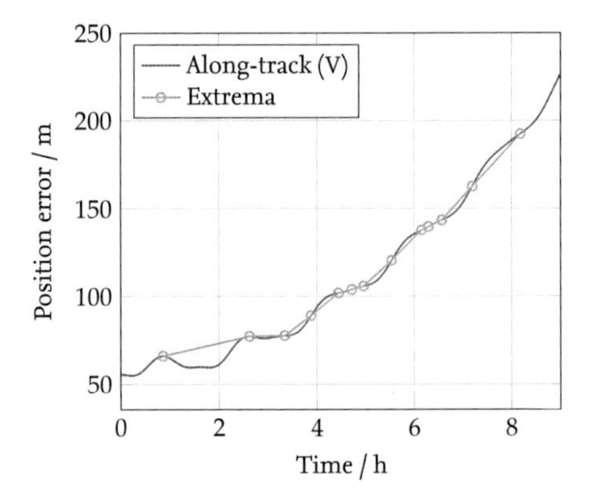

Figure 5.9.: Along-track error with **filtered** extrema being identified. Linear interpolation for adjacent extrema, shown with orange line, will be used in subsequent filter steps.

expected to contribute to an improved interpolation result of the envelope in the end. In order to evaluate which individual points need to be shifted, the following algorithm was used:

1. Perform linear interpolation for adjacent points, providing a connecting line $\lambda(t)$.

2. For each pair of adjacent points at t_i and t_{i+1}, find time $t_{max,i}$ for maximum difference between $V(t)$ and $\lambda(t)$:

$$t_{max,i} = \underset{t_i}{\operatorname{argmax}} \left(V(t) - \lambda(t)\right), \; \forall\, t \in [t_i, t_{i+1}]$$

3. If $t_{max,i} > 0$, then shift either $V(t_i)$ or $V(t_{i+1})$, depending on which one is closer to $t_{max,i}$.

By using the above formulation, there will be no shift, if $V(t)$ is always smaller than $\lambda(t)$ between two points.

The results of this algorithm are shown in Figure 5.10. The advantage of having this

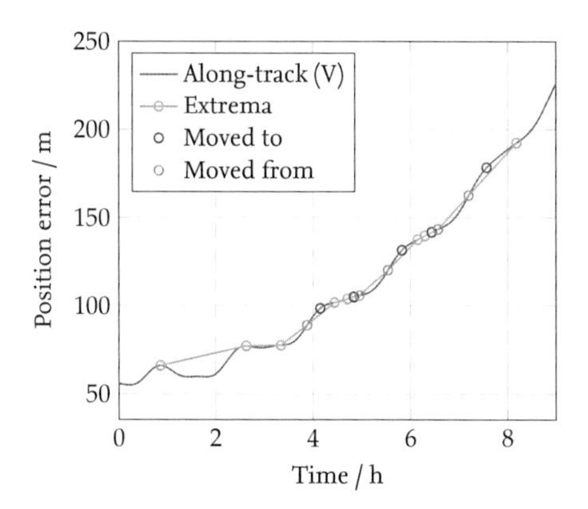

Figure 5.10.: Along-track error after **point shift filter** has been applied. Shifted points shown with red markers, including their initial values in gray.

filter in place is not really perceivable yet, but it becomes obvious after the next filter has been applied to the set of the remaining points.

Resolve clustering filter The clustering of points, as can be seen in Figure 5.10, for example at $t = 6\,\text{h}$, may put additional emphasis on the time interval comprising the cluster, to the disadvantage of the other points in the interpolation span - especially if a low-degree polynomial is used for the envelope.

The idea used to resolve the clusters is based on the median time separation M_t between two points each for the entire interpolation span. All points $i+1$ are removed, which follow with:

$$\Delta t = t_{i+1} - t_i < M_{t/2}. \tag{5.10}$$

The result of this filter is shown in Figure 5.11. It can be seen that the remaining points are now well distributed across the envisaged interpolation interval. Also, the advantage of the point shift filter from the previous step becomes clear now: If there would not have

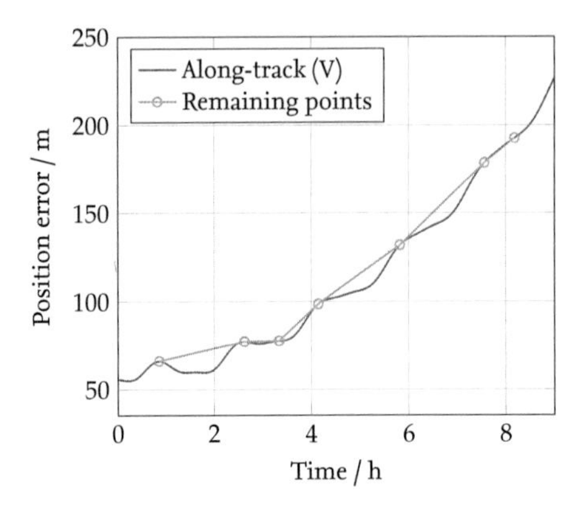

Figure 5.11.: Along-track error after **resolving clusters.**

been any shifts, connecting lines between the remaining points would, in some cases, still show intersections with $V(t)$, which is not desirable. In fact, the optimum solution, in view of the envelope computation, is to have all connecting lines being tangent to $V(t)$.

Final optimization and interpolation In principle, the interpolation can already be done on the resulting set of points shown in Figure 5.11. However, it turned out that a second run of all three previously described filters looked promising and worked out well in this example. After removing two additional points, the final result was obtained by using a 5th degree Chebyshev interpolation for the envelope. It is shown in Figure 5.12.

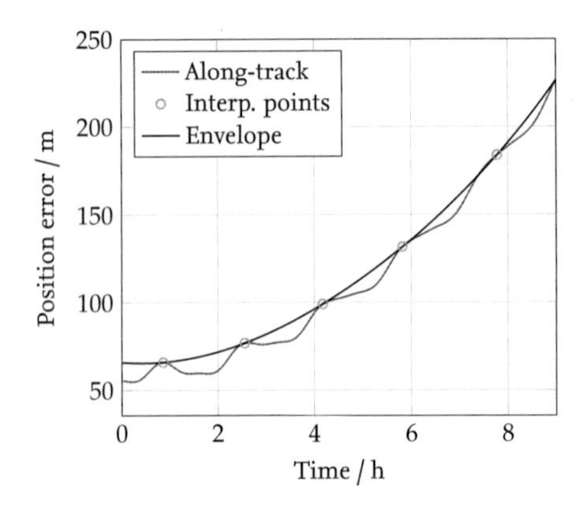

Figure 5.12.: Final result for the interpolated envelope of the along-track error.

The results were shown for one single example. An analysis for several hundred objects was performed and showed that the method was very robust, after the fit period for the envelope was set to at least five orbit revolutions. The obtained envelope interpolation results were used for probability risk computations in Section 6.2.3.

6

Applicability in the operational context

6.1. Providing orbit information via standardised data messages

The exchange of orbit information between different parties presupposes explicitly defined interfaces or data message formats. Besides the TLE data, which is a format well known and has been widely used by the community for decades now, more elaborate data messages have been defined in the recent years.

In particular, the CCSDS has developed several tailored message formats serving different purposes. For the exchange of orbital information, the CCSDS Blue Book "Orbit Data Messages" (CCSDS 502.0-B-2, 2009) defines the OEM, Orbit Mean-Element Message (OMM) and OPM, respectively. In 2012, ISO adopted this standard as ISO 26900:2012 (ISO, 2012). The OEM provides the means to exchange state vector and covariance information for different epochs as well as detailed additional information about the reference frame, object properties, etc. Also, a set of keywords (or tags for the XML realization) related to interpolation are defined. Users of this data message thus receive information on permissible time intervals in the ephemeris file that can be used for interpolation as well as on the recommended interpolation method and polynomial degree.

Being already a good option to provide Chebyshev polynomials for state and covariance information, a direct provision of the polynomial coefficients is not possible. This may, however, be overcome with the new Orbit Comprehensive Message (OCM), which is currently being prepared for implementation in a future revision of the Orbit Data Messages Blue Book. The OCM aims at providing more flexibility through the combination and extension of the OEM, OMM and OPM in a single message.

One solution could be to use the envisaged USER_DEFINED_x keyword, where x is to be replaced by any user-specified string. A combination of these keywords could thus be used to include polynomial coefficients into the OCM. Being still in a draft status, one could continue to discuss whether it makes sense to actually have dedicated keywords to directly provide polynomials for state and covariance information.

```
<ocm>
  <header>
    ...
  </header>
  <body>
    &lt;segment&gt;
      <metadata>
        ...
        <useable_start_time>...</useable..>
        <useable_stop_time>...</useable..>
        ...
      </metadata>
      <data>
        <prx1>3.456e-2</prx1>
        <pry1>7.890e-1</pry1>
        ...
        <pvz1>1.234e-5</pvz1>
        ...
        <pvzN>5.678e-6</pvzN>
      </data>
    &lt;/segment&gt;
    &lt;segment&gt;
      ...
    &lt;/segment&gt;
    ...
```

Figure 6.1.: Example of how an OCM (in XML format) might be used to provide polynomials of degree N for the radius and velocity components of the state vector.

In Figure 6.1 an example is shown for how an OCM might be used to provide polynomial coefficients. The header provides some general information, including the originator and the version of the data file. The message body contains several segments, which are defined by the start and stop time of the individual interpolation intervals. Both, begin and end epoch for each segment are identified by the *useable_start_time* and *useable_stop_time*, respectively, which are part of the metadata section within the segment. On the user's side, nothing more needs to be done than reading this file, taking care of selecting the right segment for the epoch under consideration and let the polynomials being reconstructed using a standard Chebyshev polynomial processor. However, there is still a problem to be adressed on the data center's side regarding the segmentation, as the latter may result in discontinuities, which shall be one of the identified performance issues discussed next.

6.2. Performance issues

6.2.1. Segmentation

In Section 5.2 it was explained that in order to keep the residuals to the reference trajectory within desired bounds, the polynomial degree and the granule or interval length cannot be selected independently. Therefore, in order to provide data files covering several days of an orbit, the means to perform a segmentation of the full time span into manageable time intervals are required. The subdivision of the time span into several granules and the interpolation of each of them independently can be easily accomplished and included in a data message as shown in Figure 6.1. However, it is not guaranteed that the transition of one segment to the subsequent one is continuous (and differentiable) in the different components.

An example illustrates the problem for a typical sun-synchronous orbit, shown in Figure 6.2. For this example, the Sentinel-1A orbit was used, with the granule length set to

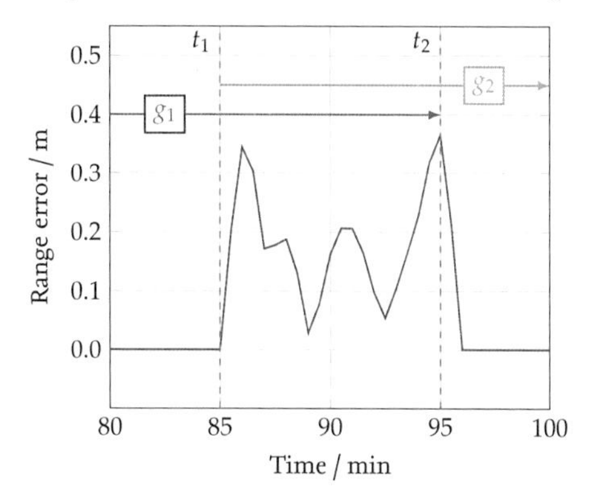

Figure 6.2.: Example for the position or range error for two overlapping interpolation intervals g_1 and g_2, showing the effect of a discontinuity of subsequent intervals. The orbit of Sentinel-1A was used, with a granule length of 90 min and a maximum accepted error of 1 m for each radius component wrt. the reference trajectory.

90 min, and the maximum allowed position error for the individual components was set to 1 m. By letting the two adjacent intervals g_1 and g_2 overlap, it is possible to compute the difference between the two polynomials for a given epoch. The overlap was selected as 10 min in this example.

As can be seen in Figure 6.2 the discontinuity caused by the segmentation process introduces an additional error resulting in an RMS of 19 cm for the residuals. In this example, this translates to 19 % of the maximum accepted error of 1 m.

One idea to overcome this problem is to post-process the first interpolation result for adjacent segments using a weight function for a pre-defined time span t_s centered at the interval transition time t_τ between g_1 and g_2. Such a weight function would give full weight to g_1 and g_2 at $t_1 = t_\tau - t_s/2$ and $t_2 = t_\tau + t_s/2$, respectively. Letting the weights of the two segments decline according to a cosine law, the following formulation can be used for the correction of the individual components x_i of the state vector:

$$x_i(t) = \frac{1}{2} \cdot (w_{i,1} + w_{i,2}) \tag{6.1}$$

$$w_{i,1} = x_{i,g_1} \cdot \left[1 + \cos\left(\frac{t - t_1}{t_s} \cdot \pi\right)\right] \tag{6.2}$$

$$w_{i,2} = x_{i,g_2} \cdot \left[1 + \cos\left(\frac{t_2 - t}{t_s} \cdot \pi\right)\right] \tag{6.3}$$

Of course, adapting the state vector components at the begin and the end of an interval of a segment will render the polynomial coefficients of the original interpolation invalid. Therefore, a second interpolation can be performed for the same segments, in order to update the polynomial coefficients, which will then correspond to the solution with a smooth and continuous transition for adjacent segments. For the same example from above, the second interpolation results are shown in Figure 6.3 for the residuals of the interpolation results with respect to the reference trajectory. Although the second inter-

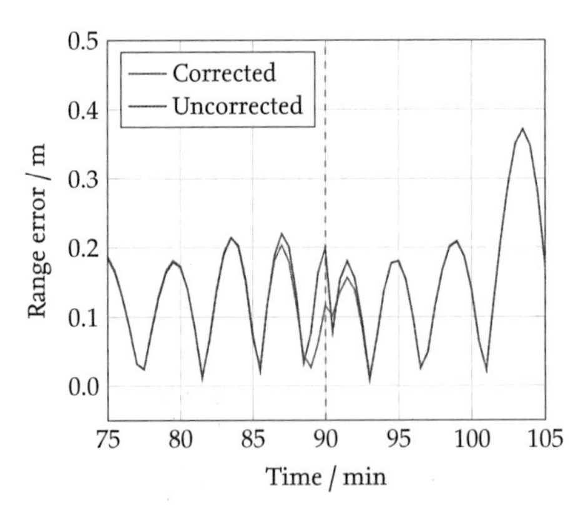

Figure 6.3.: Difference in range error between *uncorrected* (first interpolation without weight function) and *corrected* (second interpolation after weight function was applied) interpolation with the segment transition epoch being centered.

polation is computed on the set of data points from the first interpolation polynomial - as opposed to the first interpolation, which is based on the reference trajectory - it can be seen that notable differences between the two interpolations occur only in the region where the weight function was applied. However, the range error is still bounded in this example and significantly below the maximum accepted error of 1 m.

In addition, due to the denser sampling of data points at the interval edges, the Chebyshev interpolation already comes with a reduced error in this area when compared to mid-interval values. This provides additional stability for the application of the weight function.

It is also important to look at the velocity errors at the interval transition. In the context of conjunction assessment, in particular, the estimation of a collision probability is sensitive to the velocity vector, so that it has to be assured that the velocity errors in the individual components are in an acceptable regime.

The velocity vector component errors (after the second, corrected interpolation) are shown, for the same Sentinel-1A example, in Figure 6.4. For an AEL of 1 m, the velocity

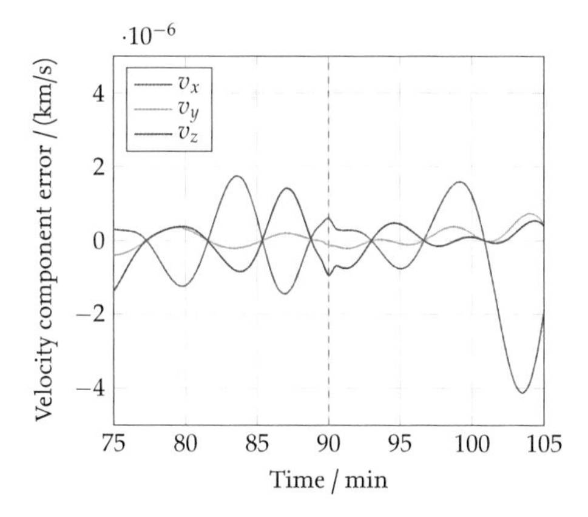

Figure 6.4.: Errors in the individual velocity components (final result after second interpolation with weight function) with the segment transition epoch being centered.

errors are on the order of a few mm s^{-1}. More importantly, the transition from the first to the second interval looks quite smooth, so that it can be concluded that also the velocity results are credible at the interval transitions.

6.2.2. Data message size and update cycles

The data compression ratio r_c shall be defined as follows:

$$r_c = \frac{\text{Ephemerides-based message size}}{\text{Polynomial-based message size}} \tag{6.4}$$

Assuming that the number of digits for each number provided in both the ephemerides-based (tabulated) as well as the polynomial-based message, is the same, one can simply derive a formula for the compression ratio, which only depends on the time interval used for the ephemerides and the coefficients-per-hour figures given in Table 5.2. The message size for the ephemerides-based message, $s_{m,eph}$, considering a k-dimensional state vector and a $n \times n$ covariance matrix (with n^* elements being stored), can be computed as:

$$s_{m,eph} = \frac{k + n^*}{\Delta\tau} \cdot \Delta t, \tag{6.5}$$

where $\Delta\tau$ is the step size of the ephemerides in the file and Δt is the time interval covered by the message. Note that only the *data* part of the message is considered, as *header* and *metadata* sections are assumed to be of the same size for both messages. For the size of the polynomial-based message, $s_{m,pol}$ one obtains:

$$s_{m,pol} = (k \cdot c_S + n^* \cdot c_C) \cdot \Delta t, \tag{6.6}$$

where c_S is the number of coefficients per hour for the state vector polynomials from Table 5.2, while c_C is the equivalent number of coefficients per hour for the covariance matrix elements. The compression ratio can now be written as:

$$r_c = \frac{k + n^*}{\Delta\tau \cdot (k \cdot c_S + n^* \cdot c_C)}. \tag{6.7}$$

For a message containing only state vector information ($n^* = 0$), Equation 6.7 can be further simplified:

$$r_{c,S} = \frac{1}{c_S \cdot \Delta\tau}. \tag{6.8}$$

With both the coefficients-per-hour ratio and the step size being in the denominator in Equation 6.8, it is clear that the compression ratio will increase for decreasing step sizes in the ephemerides-based message as well as for decreasing coefficient-per-hour ratios c_S, the latter being a function of the AEL and the orbit. An example for an AEL of 1 m is shown in Figure 6.5, also depicting typical step sizes for POE which can be accessed online. The compression ratio is well above one for LEOs and GTOs for ephemeris step sizes of a few minutes. Setting the AEL to 10 m is further improving this ratio, as shown in Figure 6.6.

Using the envelope interpolation method for the covariance matrix compression introduced in Section 5.3 would improve the situation even more. Taking the example from Section 5.3, the covariance matrix is provided with six polynomial coefficients per component for a time span of about 8 h. This corresponds to 0.075 coefficients per minute and thus in the comparison with tabulated data to a compression ratio above one for step sizes less than about 13 min. This can be easily verified by setting $k = 0$ in Equation 6.7.

It is thus likely to achieve a compression ratio above one with polynomial-based data messages. More important, however, are at least two other properties of the presented methods: firstly, polynomials provide an easy way of obtaining state vectors and uncertainties for any epoch, while messages based on tabulated ephemerides have to be propagated or interpolated to obtain intermediate values. Secondly, the proposed method of having the envelope function provided as a polynomial comes with the advantage of easily forecasting the uncertainties.

The envelope function is also useful in estimating when the next update will be required. In (Krag et al., 2010), a detailed analysis of design drivers for a space surveillance

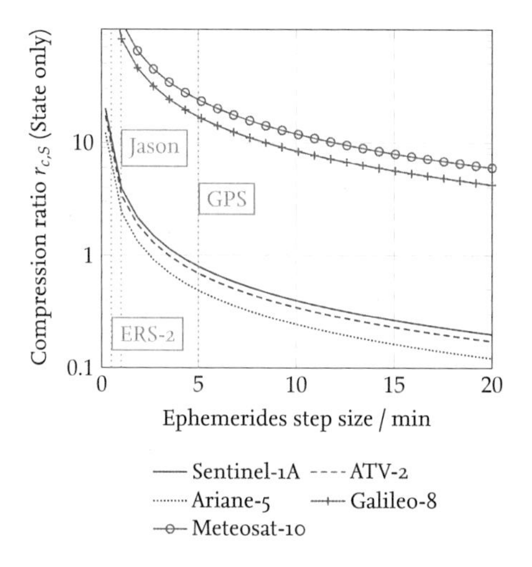

Figure 6.5.: Compression rates for messages containing only state vector data as a function of the step size between subsequent ephemerides. An AEL of 1 m was used for the polynomials. Also shown are step sizes exemplarily for POEs available online.

and tracking system was performed. The authors also provided a justified method to specify update cycles for state vectors and covariances. A so-called *covariance envelope* was introduced (not to be mixed up with the *envelope function* introduced here). The updates are supposed to occur as soon as the uncertainties are above a given threshold. Using the envelope function, one can easily determine the validity time span of a data set.

6.2.3. Covariance envelope interpolation in the operational collision avoidance context

The envelope interpolation outlined in Section 5.3 is applied to the diagonal elements only, or the state vector variances. Being the supremum function, the interpolated envelope does not affect the positive definiteness of the matrix. However, the variances will generally be greater than the propagated reference. Also, it is not possible to apply the same envelope approach to the off-diagonal matrix elements, or covariances, as the correlation coefficients show oscillations and might be positive or negative for a given point in time.

In order to find out how far one can go with reducing the required information to be stored for the covariance matrix, a study was performed on a set of CDMs. The idea was that if there is only an envelope interpolation for the diagonal elements, one could try to omit the off-diagonal elements and see, whether essential services can still be operated with variances-only information.

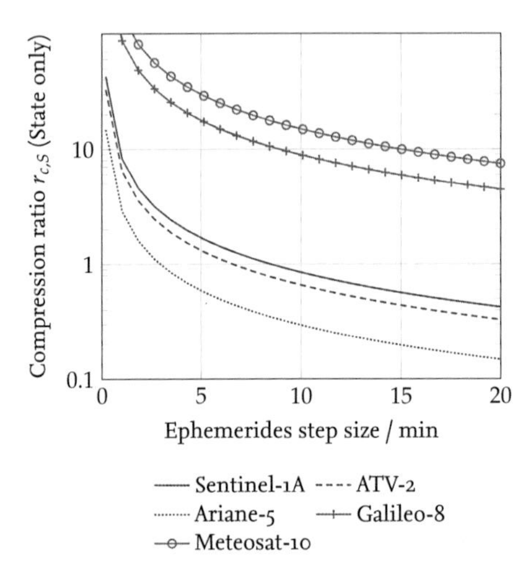

Figure 6.6.: Compression rates for messages containing only state vector data as a function of the step size between subsequent ephemerides. An AEL of 10 m was used for the polynomials. Also shown are step sizes exemplarily for POEs available online.

The information on the uncertainty in the state vector is used to compute the collision probability at the TCA. The latter is an essential input for manoeuvre decisions in the operational collision avoidance.

A common method to compute the collision probability for a close encounter is based on the following equation (e.g. Alfriend et al. (1999); Klinkrad et al. (2005):

$$p_{col} = \frac{1}{2\pi\sqrt{|\mathbf{P}_{cmb}|}} \int_A \exp\left(-\frac{1}{2}\Delta\mathbf{r}^T\mathbf{P}_{cmb}^{-1}\Delta\mathbf{r}\right) dA, \tag{6.9}$$

where $\Delta\mathbf{r}$ is the miss distance vector at TCA, $\mathbf{P}_{cmb} = \mathbf{P}_{tgt} + \mathbf{P}_{chs}$ is the projection of the combined covariance matrix of the target and chaser satellites on the so-called *B-plane* (Foster and Estes, 1992), which is a plane perpendicular to the relative velocity difference of both objects at TCA. The integration area is a circle, which encompasses the projected cross-sections of both objects and is centered at $\Delta\mathbf{r}$.

For the CDM analysis, it was especially convenient that JSpOC started sharing full 6×6 covariance matrices from January 19, 2016 with satellite owners and operators. A set of 783 CDMs received for ESA satellites in LEO was selected. All contained a full covariance matrix for both target and chaser objects, which is a prerequisite for a covariance matrix propagation and subsequent computation of an envelope and its interpolation coefficients. All selected CDMs had a miss distance less than 1000 m, in order to focus mainly on messages where non-negligible collision probability was expected.

The first analysis consisted of computing the collision probability according to Equation 6.9 at TCA for the full covariance matrix versus a covariance matrix which contained only the (original) variances on the diagonal, with all off-diagonal elements set to zero.

The second step was then to use the covariance matrix from the CDMs, propagate it for at least five orbits and then to compute the envelope function. The result of the envelope was used to obtain the value it provides at TCA and assess the collision probability.

The results are shown in Figure 6.7. It can be seen that for the first step, omitting

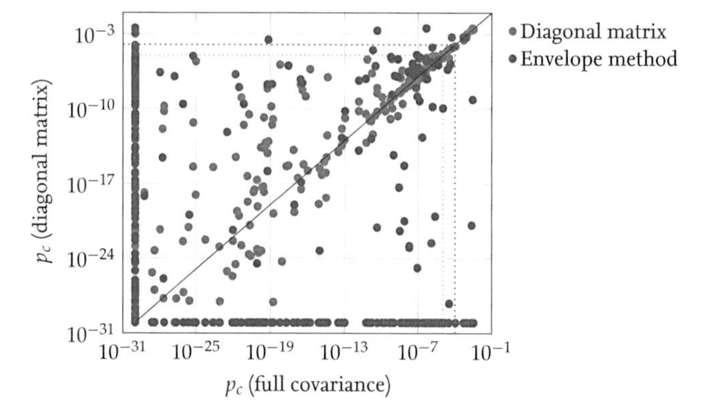

Figure 6.7.: Collision probability differences with respect to results obtained from original information in the CDM. First analysis was for using variances only (no correlation) from the CDM. The second analysis results were obtained for the envelope interpolation method. A cut-off was introduced for computed probabilities $p_{col} < 10^{-30}$. Typical operational decision thresholds at $p_{col} = 10^{-5}$ and $p_{col} = 10^{-4}$ are shown with dotted lines.

the off-diagonal terms and using only the variances from the CDMs, the differences are large for low probability values. In fact, the collision probabilities in most cases seem to be larger than for the probabilities computed from the full covariance matrix. It is interesting to note that for increasing collision probability, especially close to the range with operational relevance (e.g. $p_{col} = 10^{-5}$ or $p_{col} = 10^{-4}$ as usual decision thresholds), the results obtained by the variances-only method come close to the reference values. In fact, there was only one additional event identified with $p_{col} > 10^{-5}$, which can be regarded as one additional false alarm.

Applying the envelope interpolation method, a notable result is that events with significant collision probability often resulted in a too low estimate. One can also see that there are events which result in higher collision probability values, compared with the reference, especially for low collision probabilities.

Both effects can be explained by examining the example shown in Figure 6.8. If the variance increases, which is in general the case for the envelope method, both possibili-

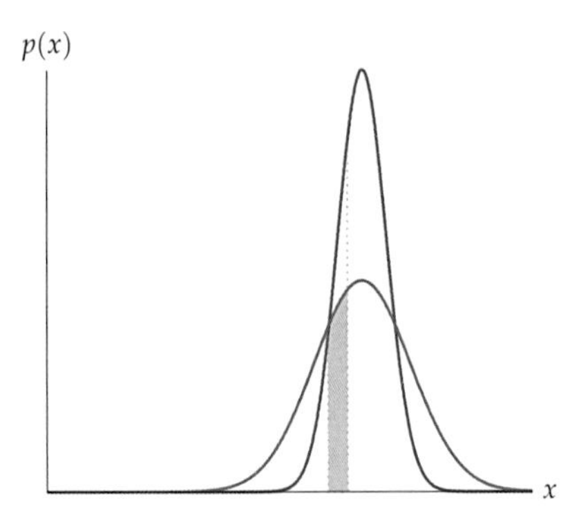

Figure 6.8.: Example showing two Gaussian Probability Density Functions (PDFs) with the same mean but different standard deviation. In a 1D-scenario, the collision probability would correspond to the integration of the PDF along the x-axis for the shown range.

ties exist: If the miss distance is large, which would correspond to a larger offset of the integration area to the mean of the PDF, the probability would increase. However, if the integration occurs close to the mean (small miss distance), one would obtain a decreased collision probability.

From an operational point of view, this means that events which would have been assessed as critical could be significantly below the decision threshold and thus might not even be considered for a more detailed screening.

The method of covariance matrix compression via its envelope function can therefore be considered as useful in applications or services where the requirements concerning the trueness of the provided covariance matrix are less strict. For example, the method solution may be beneficial in acquiring objects by deriving an associated duration for the sensor pointing from the uncertainties in the state estimate.

7 Conclusion

The aim of this thesis was to investigate specific questions related to the data exchange between a Space Surveillance and Tracking (SST) system and its external users. The different data needs and varying levels of privileges associated with the individual users, who connect to the services an SST system provides, presuppose a clearly defined data policy.

The way JSpOC used to distribute data in the frame of its collision avoidance service, is a remarkable example of how such a data policy evolves over time: while in the past satellite owners and operators had to rely on TLE orbit information to run their own conjunction predictions, JSpOC started distributing CSMs in the aftermath of the Iridium-Cosmos collision. The state vector information contained in those conjunction messages was by far more accurate than the information contained in TLE. Moreover, covariance matrix information was also provided, which allowed to assess the probabilities for each event.

The conjunction prediction service provided by JSpOC saw further improvements with the introduction of the CDM and the latest upgrade resulting in the distribution of full covariance information. Noting that CDMs have been accepted by operators as the state-of-the-art source of information for collision avoidance, clearly identifies this service as the one driving the accuracy-related requirements of an SST system. On the other hand, there are services that continue to work with less accurate TLE. For example, using TLE state vectors as apriori information for the acquisition of an object above the horizon is an adequate approach for passive optical observations.

Besides the fact that TLE come without any information on the associated uncertainties, another disadvantage is that they are based on the analytical orbit theory SGP4/SDP4: computing orbit states requires to use the same theory. The need to keep the orbit theory consistent between the data originator and the user is a source of errors often encountered in practice.

The motivation for this study was thus to investigate the possibility of deriving orbit information of predetermined accuracy from highly accurate information available in an object catalogue. This has the advantage of having only to maintain one satellite catalogue from which different data products can be derived.

Moreover, a distribution method ought to be identified which avoids having the need to keep orbit extrapolation software consistent between all parties involved in the distribution.

The following problems were addressed sequently in this thesis:

(1) Design an orbit propagation tool, based on numerical integration, for state and covariance as well as the means to assess process noise - for the purpose of being used in an SST system.

(2) Deriving from available high-accuracy data a solution with a predetermined accuracy tailored to the needs of users and services.

(3) Provide the means to establish a data message containing continuous information making use of current data message standards.

(4) Investigate a method to also provide interpolated covariance matrix information associated with the obtained state vectors of predetermined accuracy.

The numerical orbit propagation software NEPTUNE has been developed as a prerequisite for the subsequent analyses in this thesis. The software design was driven by international standards and guidelines, such as ECSS, 2008, ANSI/AIAA, 2010 or ISO, 2014. A numerical integration based on a variable-step multi-step Cowell method was selected and successfully combined with a Runge-Kutta (RK) propagation of the state transition matrix, allowing to extrapolate also the uncertainties in the state vector given by the covariance matrix.

Beyond the modelled perturbations, an approach to also account for unmodelled effects was studied. The errors introduced by the truncation of the geopotential are time-correlated and referred to as *coloured noise*. A method proposed by Nazarenko (2010) was implemented to compute those noise contributions. Together with the state vector and covariance matrix propagation, NEPTUNE offers the capability to perform the essential time update steps in the orbit determination process of an SST system.

The next step was to find a method that allows to derive an orbit with reduced accuracy from a given reference with assessable method error. Using a set of sample points as pseudo-observations of the reference trajectory, a batch least-squares approach was implemented to find a solution based on a Gauss-Newton (or Differential Correction) iteration. The required method error with respect to the reference was introduced by reducing the degree and order of the geopotential. This proved to be a promising method for orbits in the LEO region, especially after augmenting the iteration by the Levenberg-Marquardt method, which solved convergence issues for low-inclination orbits. It was shown that algorithm parameters like the fit span and number of samples can be selected in a wide range without significantly affecting the results.

As the geopotential degree can only be reduced in discrete steps, it is not possible to reach any arbitrary level of accuracy using such a method. This, however, is fully acceptable, as it can be assumed that the use-case for such a method would be to define only a couple of different solutions - for example, the current approach by JSpOC provides only one further source of information in addition to the highly-accurate SP states, namely TLE.

Different metrics can be applied to assess the deviation between the reference and the fit solution. While working with the component-wise RMS is a very intuitive and convenient way, an alternative using a more statistical description of the deviation via the Mahalanobis distance was also discussed. Obtaining a fit based on a predefined Mahalanobis distance threshold allows to take the uncertainties of the reference orbit into account: If the reference already contains large inherent errors, using a fixed threshold would lead to less degradation in the accuracy compared to an orbit with lower uncertainties.

The presented method was shown to be also applicable to orbit regions above LEO as well as to orbits with high eccentricity. With increasing altitude, the influence of the non-spherical geopotential decreases so that it could be observed for GEO altitudes, that already a 3×3 geopotential fit of the reference provides a very close fit. On the other hand, for objects with repeating ground tracks, cutting off at low degrees for the potential means missing important resonance terms. For the constellations in MEO this was shown to be advantageous in order to obtain a reduction in accuracy.

The analysis performed for high-eccentricity orbits showed that the Geopotential Adaptation Method to Bias Trajectories (GAMBIT) algorithm can be applied without any problems up to $e \approx 0.6$. For higher eccentricity, the applied approach starts showing some problems, mainly related to the sampling of pseudo-observations in equidistant time steps, which led to an under-sampling of the perigee passes with respect to the apogee region. The algorithm itself did not fail, but the required geopotential degree to stay within the specified bounds increased significantly. In addition, it was shown that in some cases the predefined accuracy bounds should not be selected in a too restrictive way.

The least-squares fit also provides a covariance matrix, which reflects the residuals with respect to the reference orbit. It was discussed how this covariance can be combined with the uncertainties inherent in the reference trajectory itself. In essence, the fit solution would be a constant offset to the propagated uncertainties of the reference solution. This renders the usual approach useless of having a covariance matrix at t_0 and propagating it to any $t \neq t_0$, but is still a valid approach, if one provides the covariance as tabulated data to cover the entire fit span. Alternatively, one can also use an interpolation of the uncertainties across the fit span and provide polynomial coefficients.

The interpolation of state vectors and their associated uncertainties as well as providing them in standardised data messages was studied as the final important aspect of this thesis. It was shown that Chebyshev polynomials are one set of polynomials where due to their orthogonality property, the orbit approximation improves by just increasing the interpolation degree. As the GAMBIT method can be compared with a low-pass filter by removing higher-degree geopotential terms, one also minimizes the risk of under-sampling the trajectory for a given number of Chebyshev nodes. Besides the demonstrated advantage of providing data messages of reduced size for a given time interval, it has to be emphasized that an interpolation solution does not require a propagation on the user's side any longer. The user can reconstruct the orbit out of the data messages in exactly the same way the data distributor would do - no need to distribute an orbit extrapolation software alongside with the data, as is the case for TLE. Chebyshev polynomials are already successfully distributed for planetary orbits by JPL, while in this thesis it was shown that they can even be applied down to orbits significantly affected by drag without introducing additional errors greater than 1 m.

The covariance matrix can also be provided via polynomial coefficients. While one can find methods in literature to interpolate the eigenvectors and their rotation, an alternative approach was studied in this thesis: For the variances in the object-centered reference frame (UVW), an envelope function was computed first. It was then interpolated with a low-degree polynomial, which results in high compression ratios compared to tabulated data. However, this is not possible for the correlation coefficients, as some of them are oscillating and may be either positive or negative depending on the position along the orbit. This means that only the variances would be provided in a data message and information on correlation between the components would be lost.

It was shown, that this negatively affects the collision avoidance service, while one can still imagine services, where such an envelope-based approach could be beneficial. For example, to obtain an initial guess for the search region to acquire an object, where conservative estimates on the variances might be advantageous.

In conclusion, the methods investigated in this thesis allow for providing orbit information continuous in time with different levels of accuracy derived from a reference orbit in an object catalogue. The modification of the orbit by a least-squares fit with an adapted geopotential ensures that also the reference state vector is adapted. Performing this modification step followed by a Chebyshev interpolation with sufficiently high degree, thus retaining the accuracy through the interpolation step, is essential: the alternative would be to have a predetermined accuracy directly via reducing the degree of the polynomial fit - but this has a significant drawback, as polynomials will always match the original trajectory at several points across the fit span. Moreover, one would have to deal with undersampling issues, as the high frequencies due to the geopotential are still present and likely to be interpolated with a polynomial of too low degree.

If the recovery of the original information, or the high-accuracy orbit from the catalogue, is not possible from the provided solution, this gives rise to the possibility of giving groups of users different privileges in terms of how accurate the information is they can obtain from the catalogue. Moreover, this would also be a way of commercialising the data distribution by offering different categories of orbit information with varying accuracy.

It would be an interesting aspect for future study to investigate whether or not it is possible to restore the original information from a least-squares fit. In principle, this could work, if the force model, that was used to generate the reference trajectory, is known to some extent. Then, one could solve for the remaining parameters by the following iteration:

1. Fit a trajectory with full force model on the input orbit

2. Do another fit on the obtained solution, now with the adapted geopotential

3. Compare with the input orbit

4. Take the deviation as an input to update the trajectory in the first step and repeat the process until both the input orbit and the solution from step 2 match

Whether or not such an approach works in the end, is difficult to answer, as the parameter space can be very large. Also, the author is not aware of any publication featuring such a method, for example to obtain the original SP states from available TLE.

An important aspect for future studies is to investigate how alternative techniques for uncertainty propagation affect the methods presented in this thesis. Especially for the covariance envelope interpolation, the results could be completely different, if a non-linear propagation is introduced.

The NEPTUNE software developed in the frame of the Networking/Partnering Initiative (ESA research programme with industry and academia) (NPI) can be further developed to be used in the orbit determination context. In that respect, the next step would be to expand it to process observations and the associated partial derivatives. In fact, the evolution of the software in that direction would be in line with the goals of the NPI to prepare academia for ESA programmes - in this case the SSA programme.

Finally, with standardised data messages evolving, especially those recommended by CCSDS, it is worth contributing to those efforts and promote the option of having polynomials in those messages, like the example outlined in this thesis for a possible implementation in the OCM.

References

M. Abramowitz and I. A. Stegun. *Handbook of Mathematical Functions with Formulas, Graphs, and Mathematical Tables.* Dover, 9th Dover printing, 10th GPO printing edition, 1964.

S. Aida and M. Kirschner. Collision Risk Assessment and Operational Experiences for LEO Satellites at GSOC. In 22nd *International Symposium on Space Flight Dynamics (ISSFD), São José dos Campos, Brazil,* 2011.

K. Aksnes. Short-period and long-period perturbations of a spherical satellite due to direct solar radiation. *Celestial Mechanics,* 13(1):89–104, 1976.

S. Alfano et al. Orbital covariance interpolation. 2004.

K. T. Alfriend, M. R. Akella, J. Frisbee, J. L. Foster, D.-J. Lee, and M. Wilkins. Probability of collision error analysis. *Space Debris,* 1(1):21–35, 1999.

F. Allahdadi, I. Rongier, P. Wilde, and T. Sgobba. *Safety Design for Space Operations.* Elsevier Science, 2013.

American Institute of Aeronautics and Astronautics. Guide to Reference and Standard Atmosphere Models. Technical Report AIAA G-003C-2010, Reston, VA, 2010.

American National Standards Institute (ANSI). Astrodynamics - Propagation Specifications, Technical Definitions, and Recommended Practices. Technical Report ANSI/AIAA S-131-2010, published by American Institute of Aeronautics and Astronautics, Reston, VA, August 2010.

P. D. Anz-Meador. International guidelines for the preservation of space as a unique resource. *Online Journal of Space Communication,* 6, 2004.

R. Armellin, P. Di Lizia, F. Bernelli Zazzera, and M. Berz. Asteroid Close Encounters Characterization Using Differential Algebra: the Case of Apophis. *Celestial Mechanics and Dynamical Astronomy,* 107(4):451–470, 2010.

Ball Aerospace & Technologies Corp. SBSS Space Vehicle, Jan. 2015. URL `http://www.ballaerospace.com/file/media/SBSS%2006_10.pdf`.

F. Barlier, F. Mignard, M. Carpino, and A. Milani. Nongravitational perturbations on the semimajor axis of LAGEOS. *Annales Geophysicae,* 4:193–210, 1986.

M. M. Berry. *A variable-step double-integration multi-step integrator.* PhD thesis, Virginia State University, 2004.

M. Berz, editor. *Modern Map Methods in Particle Beam Physics, Vol. 108*. Academic Press, 1999.

D. Bird. Sharing Space Situational Awareness Data. In *Proceedings of the Advanced Maui Optical and Space Surveillance Technologies Conference, AMOS, Maui, Hawaii*, 2010.

N. Bobrinsky. The European SSA Preparatory Programme. In *5th meeting of the PECS committee, ESRIN, Italy*, 2009.

V. F. Boikov, G. N. Makhonin, A. V. Testov, Z. N. Khutorovsky, and A. N. Shogin. Prediction Procedures Used in Satellite Catalog Maintenance. *Journal of Guidance, Control, and Dynamics*, 32(4):1179–1199, July 2009.

B. Bowman. Personal e-mail correspondence, March 2014.

V. A. Brumberg. *Analytical Techniques of Celestial Mechanics*. Springer Verlag Berlin Heidelberg New York, 1995.

N. Capitaine, P. T. Wallace, and J. Chapront. Expressions for IAU 2000 precession quantities. *Astronomy & Astrophysics*, 412:567–586, 2003.

S. Casotto. *Nominal ocean tide models for TOPEX precise orbit determination*. PhD Thesis, University of Texas at Austin, 1989.

V. Coppola, J. H. Seago, and D. A. Vallado. The IAU 2000A and IAU 2006 Precession-Nutation Theories and their Implementation, Paper AAS 09-159. In *AAS/AIAA Astrodynamics Specialist Conference*, 2011.

P. H. Cowell and A. C. Crommelin. Investigation of the motion of Halley's comet from 1759 to 1910. *Appendix to Greenwich Observations for 1909*, pages 1–84, 1909.

A. Deprit, H. Pickard, and W. Poplarchek. Compression of ephemerides by discrete Chebyshev approximations. *Navigation*, 26(1):1–11, 1979.

P. Di Lizia, R. Armellin, and M. Lavagna. Application of High Order Expansions of Two-Point Boundary Value Problems to Astrodynamics. *Celestial Mechanics and Dynamical Astronomy*, 102(4):355–375, 2008.

D. P. Drob, J. T. Emmert, G. Crowley, J. M. Picone, G. G. Shepherd, W. Skinner, P. Hays, R. J. Niciejewski, M. Larsen, C. Y. She, et al. An empirical model of the Earth's horizontal wind fields: HWM07. *Journal of Geophysical Research: Space Physics (1978–2012)*, 113(A12), 2008.

R. Eanes and S. Bettadpur. The CSR 3.0 Global Ocean Tide Model, CSR-TM-95-06. Technical report, Center for Space Research, University of Texas, Austin, TX, 1995.

P. R. Escobal and R. A. Robertson. Lunar eclipse of a satellite of the Earth. *Journal of Spacecraft and Rockets*, 4:538–540, 1967.

European Cooperation for Space Standardization (ECSS). Space engineering - Space environment. Technical Report ECSS-E-ST-10-04C, Noordwijk, The Netherlands, November 2008.

European Space Agency Council. *Space Situational Awareness Preparatory Programme Proposal.* ESA/C(2008)142, 2008.

T. Flohrer, H. Krag, and H. Klinkrad. Assessment and Categorization of TLE Orbit Errors for the US SSN Catalogue. In *Proceedings of the Advanced Maui Optical and Space Surveillance Technologies Conference, AMOS, Maui, Hawaii*, 2008.

C. Förste, R. Schmidt, R. Stubenvoll, F. Flechtner, U. Meyer, R. König, H. Neumayer, R. Biancale, J. Lemoine, S. Bruinsma, S. Loyer, and F. Barthelmes. The GFZ/GRGS satellite and combined gravity field models EIGENGL04S1 and EIGEN-GL04C. *Journal of Geodesy*, 82(6):331 – 346, 2008.

J. L. Foster and H. S. Estes. A parametric analysis of orbital debris collision probability and maneuver rate for space vehicles. *NASA JSC*, 25898, 1992.

B. R. Frieden. *Physics from Fisher information: a unification.* Cambridge University Press, 1998.

E. V. Gavrilin. *Jepoha klassicheskoj raketno-kosmicheskoj oborony.* Tehnosfera, Moscow, 2008.

A. Gelb. *Applied Optimal Estimation.* MIT Press, 1974.

J. W. Gibbs. Fourier's series. *Nature*, 59:200, 1898.

A. Gil, J. Segura, and N. M. Temme. *Numerical Methods for Special Functions.* Society for Industrial and Applied Mathematics, Philadelphia, PA, USA, first edition, 2007.

D. Giza, P. Singla, and M. Jah. An approach for nonlinear uncertainty propagation: Application to orbital mechanics. In *AIAA Guidance, Navigation, and Control Conference, Chicago IL*, pages 1–19, 2009.

GlobalSecurity.org. Russian Space Surveillance System, Jan. 2015. URL `http://www.globalsecurity.org/space/world/russia/space-surveillance.htm`.

M. D. Hejduk, S. J. Casali, D. A. Cappellucci, N. L. Ericson, and D. E. Snow. A catalogue-wide implementation of general perturbations orbit determination extrapolated from higher order orbital theory solutions. In *Spaceflight Mechanics Conference, Kauai, HI, USA*, 2013.

F. R. Hoots and R. L. Roehrich. Spacetrack Report No. 3: Models for propagation of NORAD element sets. Technical report, Aerospace Defense Center, Peterson Air Force Base, 1980.

F. R. Hoots, P. W. Schumacher, and R. A. Glover. History of Analytical Orbit Modeling in the U.S. Space Surveillance System. *Journal of Guidance, Control and Dynamics*, 27(2), 2004.

M. Horemuž and J. V. Andersson. Polynomial interpolation of GPS satellite coordinates. *GPS solutions*, 10(1):67–72, 2006.

A. Horstmann, V. Braun, and H. Klinkrad. Performance of variable step numerical integration across eclipse boundary crossings for HAMR objects. In *AAS/AIAA Astrodynamics Specialist Conference, Vail, CO, AAS 15-506*, 2015.

International DORIS Service. Website (last access on 2015-01-15). http://ids-doris.org/system/37-poe/51-poe-description.html.

International Organization for Standardization (ISO). Accuracy (trueness and precision) of measurement methods and results – Part 1: General principles and definitions. Technical Report ISO 5725-1:1994, Geneva, Switzerland, December 1994.

International Organization for Standardization (ISO). Data elements and interchange formats - Information interchange - Representation of dates and times. Technical Report ISO 8601:2004, Geneva, Switzerland, December 2004.

International Organization for Standardization (ISO). Space data and information transfer systems - Orbit data messages. Technical Report ISO 26900:2012, Geneva, Switzerland, July 2012.

International Organization for Standardization (ISO). Space systems - Orbit determination and estimation - Process for describing techniques. Technical Report ISO 11233:2014, Geneva, Switzerland, April 2014.

B. A. Jones, A. Doostan, and G. H. Born. Nonlinear Propagation of Orbit Uncertainty Using Non-Intrusive Polynomial Chaos. *Journal of Guidance, Control and Dynamics*, 36(2): 430–444, 2013.

E. Kahr, O. Montenbruck, and K. P. G. O'Keefe. Estimation and Analysis of Two-Line Elements for Small Satellites. *Journal of Spacecraft and Rockets*, 50(2):433–439, Mar. 2013.

W. M. Kaula. *Theory of Satellite Geodesy: Applications of Satellites to Geodesy*. Dover Earth Science Series. Dover Publications (re-print from 1966), 2000.

T. Kelecy, E. Baker, P. Seitzer, T. Payne, and R. Thurston. Prediction and tracking analysis of a class of high area-to-mass ratio debris objects in geosynchronous orbit. In *AMOS Technical Conference, Wailea, Hawaii*, 2008.

T. S. Kelso. Validation of SGP4 and ICD-GPS-200 against GPS Precision Ephemerides. In *AIAA/AAS Astrodynamics Specialist Conference, Paper AIAA-2007-127, Sedona, AZ*, 2007.

Z. N. Khutorovsky. Low-Perigee Satellite Catalog Maintenance: Issues of Methodology. In *Second European Conference on Space Debris, ESOC, Darmstadt, Germany*, 2007.

H. Klinkrad. *Space Debris: Models and Risk Analysis*. Springer Praxis Books. Springer, 2006.

H. Klinkrad. Space Debris Mitigation Activities at ESA in 2013. In *Fifty-first session of the UNCOPUOS Scientific and Technical Subcommittee, Vienna, Austria*, 2014.

H. Klinkrad and B. Fritsche. Orbit and attitude perturbations due to aerodynamics and radiation pressure. In *ESA Workshop on Space Weather, Noordwijk, ESTEC*, 1998.

H. Klinkrad, J. Alarcon, and N. Sanchez. Collision avoidance for operational ESA satellites. In *4th European Conference on Space Debris*, volume 587, page 509, 2005.

P. Knocke. Earth radiation pressure effects on satellites, 1989.

G. Kopp and J. L. Lean. A new, lower value of total solar irradiance: Evidence and climate significance. *Geophysical Research Letters*, 38(1), 2011.

H. Krag, T. Flohrer, H. Klinkrad, and E. Fletcher. The European Surveillance and Tracking System - Services and Design Drivers. In *Proceedings of the SpaceOps Conference, Huntsville, Alabama*, 2010.

G. T. Kremer, R. M. Kalafus, P. V. Loomis, and J. C. Reynolds. The effect of selective availability on differential GPS corrections. *Navigation*, 37(1):39–52, 1990.

F. T. Krogh. Changing stepsize in the integration of differential equations using modified divided differences. Proceedings of the conference on the Numerical Solution of Ordinary Differential Equations, October 1972, Lecture Notes in Math., vol. 362, pp. 22-71. Springer-Verlag New York, 1974.

G. Lange. Characteristics of the IAU-based GCRF/ITRF transformations in view of orbit propagation applications. Research thesis, Institute of Aerospace Systems, TU Braunschweig, 2014.

D.-J. Lee and K. T. Alfriend. Sigma Point Filtering for Sequential Orbit Estimation and Prediction. *Journal of Spacecraft and Rockets*, 44(2):388–398, 2007.

F. G. Lemoine, S. C. Kenyon, J. K. Factor, R. G. Trimmer, N. K. Pavlis, D. S. Chinn, C. M. Cox, S. M. Klosko, S. B. Luthcke, M. H. Torrence, et al. The development of the joint nasa gsfc and the national imagery and mapping agency (nima) geopotential model egm96. *NASA/TP-1998-206861*, 1998.

T. Letellier, F. Lyard, and F. Lefebre. The new global tidal solution: FES2004. In *Ocean Surface Topography Science Team Meeting, St. Petersburg, Florida*, 2004.

K. Levenberg. A method for the solution of certain problems in least squares. *Quarterly of Applied Mathematics*, 2:164–168, 1944.

C. Levit and W. Marshall. Improved orbit predictions using two-line elements. *Advances in Space Research*, 47(7):1107–1115, 2011.

A. C. Long, J. O. Cappellari, C. E. Velez, and A. J. Fuchs. Goddard Trajectory Determination System (GTDS), Mathematical Theory, Revision 1. Technical report, 1989.

J. B. Lundberg. Multistep integration formulas for the numerical integration of the satellite problem. Technical report, Center for Space Research, University of Texas, Austin, TX, 1981.

J. B. Lundberg. Computational errors and their control in the determination of satellites orbits, 1985.

J. B. Lundberg, M. R. Feulner, P. A. M. Abusali, and C. S. Ho. Improving the numerical integration solution of satellite orbits in the presence of solar radiation pressure using modified back differences. In *Proceedings of the 1st AAS/AIAA Annual Spaceflight Mechanics Meeting, Houston*, 1991.

S. B. Luthcke, N. Zelensky, D. D. Rowlands, F. G. Lemoine, and T. A. Williams. The 1-cm orbit: Jason-1 precision orbit determination using GPS, SLR, DORIS, and altimeter data. *Marine Geodesy*, 26(3–4), 2003.

B. Luzum and G. Petit. IERS Conventions (2010). IERS Technical Note No. 36, International Earth Rotation and Reference Systems Service, Verlag des Bundesamts für Kartographie und Geodäsie, Frankfurt am Main, 2010.

K. Madsen, H. B. Nielsen, and O. Tingleff, editors. *Methods for Non-Linear Least Squares Problems*. Informatics and Mathematical Modelling, Technical University of Denmark, 2 edition, 2004.

P. C. Mahalanobis. On the generalized distance in statistics. *Proceedings of the National Institute of Sciences (Calcutta)*, 2:49–55, 1936.

M. J. Matney, P. Anz-Meador, and J. L. Foster. Covariance correlations in collision avoidance probability calculations. *Advances in Space Research*, 34(5):1109–1114, 2004.

P. S. Maybeck. *Stochastic Models, Estimation, and Control*, volume 1. Academic Press, Inc., 1979.

D. D. McCarthy. IERS Conventions (1992). IERS Technical Note No. 21, Central Bureau of IERS - Observatoire de Paris, 1996.

D. McKissock. Joint Space Operations Center Space Surveillance Mission. In *Briefing at the SpaceOps Conference, Daejeon, South Korea*, 2016.

O. Montenbruck and E. Gill. *Satellite Orbits*. Springer Verlag Berlin Heidelberg New York, 2000.

National Aeronautics and Space Administration (NASA). Jason 1 launch (press kit), December 2001.

A. I. Nazarenko. *Errors of Forecasting of Satellite Motion in the Earth Gravitational Field*. Moscow, Institute of Space Research, RAN, 2010.

H. B. Nielsen. Damping Parameter in Marquardt's Method. Technical Report IMM-REP-1999-05, Informatics and Mathematical Modelling, Technical University of Denmark, 1999.

D. Oltrogge and T. S. Kelso. Ephemeris Requirements for Space Situational Awareness. In *AAS/AIAA Space Flight Mechanics Conference, New Orleans, LA, AAS 11-151*, 2011.

Peterson Air Force Base. 20th Space Control Squadron - Fact Sheet, Oct. 2016. URL `http://www.peterson.af.mil/Portals/15/documents/AboutUs/AFD-080219-097.doc`.

J. M. Picone, A. E. Hedin, D. P. Drob, and A. C. Aikin. NRLMSISE-00 empirical model of the atmosphere: Statistical comparisons and scientific issues. *Journal of Geophysical Research*, 107(A12), 2002.

A. Riccardi, C. Tardioli, and M. Vasile. An Intrusive Approach to Uncertainty Propagation in Orbital Mechanics Based on Tchebycheff Polynomial Algebra. In *AAS/AIAA Astrodynamics Specialist Conference, Vail, Colorado*, 2015.

J. Ries, J. Bordi, and T. Urban. Precise orbits for ERS-2 using laser ranging and PRARE tracking, CSR-TM-99-03. Technical report, Center for Space Research, University of Texas, Austin, TX, 1999.

C. Runge. Über empirische Funktionen und die Interpolation zwischen äquidistanten Ordinaten. *Zeitschrift für Mathematik und Physik*, 46:224–243, 1901.

N. Sánchez-Ortiz, J. Correia de Oliveira, R. Domínguez González, E. Parrilla Endrino, J. Gelhaus, V. Braun, C. Kebschull, and H. Krag. Study Note on WP1000 (ARES upgrade). Technical report, European Space Agency, Document ID ILR/DRAMA/SN1000, 2013.

M. Schenewerk. A brief review of basic GPS orbit interpolation strategies. *GPS solutions*, 6(4):265–267, 2003.

P. W. Schumacher and F. R. Hoots. Evolution of the NAVSPACECOM Catalog Processing System Using Special Perturbations. In *Proceedings of the Fourth US/Russian Space Surveillance Workshop, US Naval Observatory, Washington, DC*, 2000.

A. M. Segerman and S. L. Coffey. Ephemeris compression using multiple fourier series. *The Journal of the astronautical sciences*, 46(4):343–359, 1998.

L. Sehnal. The Earth Albedo Model in Spherical Harmonics. *Bulletin of the Astronomical Institution of Czechoslovakia*, 30(4):199–204, 1979.

P. K. Seidelmann. *Explanatory Supplement to the Astronomical Almanac*. United States Naval Observatory, Nautical Almanac Office, University Science Books, 2006.

L. F. Shampine and M. K. Gordon. *Computer solution of ordinary differential equations: the initial value problem*. Freeman, 1975.

SpaceNews. Editorial - For Air Force and ComSpOC, Things are Looking Up, Oct. 2015. URL http://spacenews.com/editorial-for-air-force-and-comspoc-things-are-looking-up/.

V. K. Srivastava, J. Kumar, S. Kulshrestha, A. Srivastava, M. K. Bhaskar, B. S. Kushvah, P. Shiggavi, and D. A. Vallado. Lunar shadow eclipse prediction models for the Earth orbiting spacecraft: Comparison and application to LEO and GEO spacecraft. *Acta Astronautica*, 110:206 – 213, 2015.

K. F. Sundman. Mémoire sur le problème des trois corps. *Acta mathematica*, 36(1):105–179, 1913.

S. Tanygin. Efficient covariance interpolation using blending of approximate covariance propagations. *The Journal of the Astronautical Sciences*, 61(1):107–132, 2014.

B. D. Tapley, B. E. Schutz, and G. H. Born. *Statistical Orbit Determination*. Elsevier Academic Press, 2004.

The Consultative Committee for Space Data Systems (CCSDS). Orbit Data Messages, Recommended Standard. Technical Report CCSDS 502.0-B-2, Blue Book, Washington, DC, USA, November 2009.

The Consultative Committee for Space Data Systems (CCSDS). XML Specification for Navigation Data Messages, Recommended Standard. Technical Report CCSDS 505.0-B-1, Blue Book, Washington, DC, USA, December 2010.

The Consultative Committee for Space Data Systems (CCSDS). Conjunction Data Message, Recommended Standard. Technical Report CCSDS 508.0-B-1, Blue Book, Washington, DC, USA, June 2013.

U.S. Strategic Command. USSTRATCOM Space Control and Space Surveillance, Jan. 2015. URL http://www.stratcom.mil/factsheets/11/Space_Control_and_Space_Surveillance/.

J.-L. Valero, S. Albani, B. Gallardo, J. Matute, and A. O'Dwyer. Application of use case scenarios to support the development of a future SSA governance and data policy in Europe. In *Sixth IAASS International Space Safety Conference, Montreal, Canada*, 2013.

D. Vallado, P. Crawford, R. Hujsak, and T. S. Kelso. Revisiting Spacetrack Report #3. In *AIAA/AAS Astrodynamics Specialist Conference, Paper AIAA-2006-6753, Keystone, CO*, 2006a.

D. A. Vallado and P. J. Cefola. Two-Line Element Sets - Practice and Use. In *63rd International Astronautical Congress, Naples, Italy*, 2012.

D. A. Vallado and T. S. Kelso. Using EOP and Space Weather Data for Satellite Operations. In *AAS/AIAA Space Flight Mechanics Conference, Lake Tahoe, CA, AAS 05-406*, 2005.

D. A. Vallado and W. D. McClain. *Fundamentals of Astrodynamics and Applications (Fourth Edition).* Microcosm Press, 2013.

D. A. Vallado, J. H. Seago, and P. K. Seidelmann. Implementation Issues Surrounding the New IAU Reference Systems for Astrodynamics. In *AAS/AIAA Space Flight Mechanics Conference, Tampa, FL, AAS 06-134*, 2006b.

D. Vokrouhlicky, P. Farinella, and F. Mignard. Solar Radiation Pressure Perturbations for Earth Satellites - I. A complete theory including penumbra transitions. *Astronomy and Astrophysics*, 280:295–312, 1993.

D. Vokrouhlicky, P. Farinella, and F. Mignard. Solar Radiation Pressure Perturbations for Earth Satellites - III. Global atmospheric phenomena and the albedo effect. *Astronomy and Astrophysics*, 290:324–334, 1994.

D. Vokrouhlicky, P. Farinella, and F. Mignard. Solar Radiation Pressure Perturbations for Earth Satellites - IV. Effects of the Earth's Polar Flattening on the Shadow Structure and the Penumbra Transitions. *Astronomy and Astrophysics*, 307:635–644, 1996.

P. T. Wallace and N. Capitaine. Precession-Nutation Procedures Consistent with IAU 2006 Resolutions. *Astronomy & Astrophysics*, 985:981–985, 2006.

N. Wiener. The homogeneous chaos. *American Journal of Mathematics*, 60(4):897–936, 1938.

M. P. Wilkins, K. T. Alfriend, S. L. Coffey, and A. M. Segerman. Transitioning from a General Perturbations to a Special Perturbations Space Object Catalog. In *AIAA/AAS Astrodynamics Specialist Conference, Paper AIAA-2000-4238, Denver, CO*, 2000.

J. Woodburn and S. Tanygin. Position covariance visualization. *AIAA paper*, 4985, 2002.

J. R. Wright. Sequential Orbit Determination with Auto-Correlated Gravity Model Errors. *Journal of Guidance and Control*, 4(3):304–309, 1981.

J. R. Wright, J. Woodburn, S. Truong, and W. Chuba. Orbit Gravity Error Covariance. In *18th AAS/AIAA Space Flight Mechanics Meeting, Galveston, Texas*, 2008.

List of Figures

List of Tables

List of Symbols

List of Symbols

A	Area	m^2
$\mathbf{A}$	System matrix	
a	Albedo	
$\mathbf{a}$	Acceleration vector	km/s^2
a	Magnitude of acceleration vector	km/s^2
a	Semi-major axis	km
a_a	Annual wobble	rad
a_c	Chandler wobble	rad
A_p	Planetary geomagnetic activity index	
AU	Astronomical unit, 1 AU = $1.496 \cdot 10^{11}$; m	m
B	Ballistic coefficient	m^2/kg
$\mathbf{B}$	Input matrix	
b	Storage size	bytes
b	Constant coefficient	
C	Spherical harmonics cosine coefficient	
$\mathbf{C}$	Variation of constants vector	
c	Constant coefficient	
c	Polynomial coefficient	
$\mathbf{c}$	Vector of constants	
c	Covariance matrix element	
c	Speed of light, $c = 299\,792\,458\,\mathrm{m\,s^{-1}}$	
c_A	Absorption coefficient	
c_D	Drag coefficient	
$c_{R,d}$	Diffuse reflectivity coefficient	
$c_{R,s}$	Specular reflectivity coefficient	
c_R	Solar radiation pressure coefficient	
EPS	Maximum error	
ERK_D	Approximated Error	
E	Expected value	
e	Eccentricity	
e	Emissivity	
$\mathbf{e}$	Unit vector	
F	Solar flux	$W/m^2/Hz$

$\mathbf{F}$	Force vector	N
$\mathbf{F}$	Partial derivative matrix	
F	Mean longitude of the Moon minus the mean longitude of the ascending node of the Moon (Delaunay variable)	rad
f	Specific force vector	N/kg
$\mathbf{f}$	Vector function	
f	Function	
$\mathbf{g}$	Vector function	
G	Constant of gravitation, $G \approx 6.673\,84 \times 10^{11}\,\mathrm{m^3\,kg^{-1}\,s^{-2}}$	
g	Weighting factor for acceleration differences	
g'	Weighting factor for acceleration differences	
H	Hermite polynomial	
$\mathbf{H}$	Output matrix	
h	Altitude	km
h	Step size	s
$\mathbf{I}$	Identity matrix	
i	Inclination	
$\mathbf{J}$	Jacobian matrix	
J	Performance index	
K	Albedo and emissivity function	
K	Correlation function	
k	Normalized correlation function	
k	Order of interpolating polynomial	
k	Upper limit of summation	
k	Love number	
K_p	Planetary index for geomagnetic activity	
k'_n	Load deformation coefficient of degree n	
L	Cholesky decomposition of covariance matrix	
L	Model for the prediction of the performance index	
$\mathbf{le}$	Local error vector	
le	Local error magnitude	
l	Counter for natural numbers	
l	Mean anomaly of the Moon (Delaunay variable)	rad
l'	Mean anomaly of the Sun (Delaunay variable)	rad
M	Manoeuvre	
M	Mean anomaly	
M	Mass of a celestial body	kg
M	Median	
m	Order of geopotential	
m	Mass	kg
m	Counter for natural numbers	

m	Number of observations	
N	Polynomial degree	
$\mathbf{N}$	Nutation matrix	
n	Degree of geopotential	
n	Counter for natural numbers	
n	Dimension of state vector	
$\mathbf{n}_0$	Surface normal direction vector	
$\mathbf{P}$	Covariance matrix	
P	Associated Legendre functions, typically with indices n and m	
$\mathbf{P}$	Precession matrix	
P	Solar radiation pressure	N/m^2
p	Dimension of observation vector	
$\mathbf{P}$	Polynomial vector	
P	Polynomial	
p	Probability density function	
p_a	General accumulated precession in longitude	rad
p_{col}	Collision probability	
$\mathbf{Q}$	Precession-nutation matrix according to CIP approach	
q	Counter for natural numbers	
$\mathbf{Q}_{uu}$	Second moment of the process noise matrix	
$\mathbf{Q}_{xu}$	State and process noise cross-correlation matrix	
$\mathbf{R}$	Earth rotation matrix	
R	Radius of a celestial body	
$\ddot{\mathbf{r}}$	Radial acceleration vector	
$\dot{\mathbf{r}}$	Radial velocity vector	
r	Ratio	
$\dot{r}$	Magnitude of velocity vector	km/s
r	Magnitude of radius vector	km
$\mathbf{r}$	Radius vector	km
S	Spherical harmonics sine coefficient	
s	CIO locator	
s	Size	
$\mathbf{s}$	Distance vector	
$\mathbf{s}_0$	Sun normal direction vector	
s'	TIO locator	
T	Chebyshev polynomial of first kind	
$\mathbf{T}$	Coordinate transformation rotation matrix	
T	Orbital period	s
T	Manoeuvre transition phase	
t	General variable	
t	Time	s

θ	Precession angle (rotation about intermediate negative y-axis)	
U	Chebyshev polynomial of second kind	
U	Potential function	
$\mathbf{u}$	System noise vector	
u	Argument of true latitude	rad
$\mathbf{v}$	Velocity vector	km/s
v	Magnitude of velocity	km/s
$\mathbf{W}$	Polar motion rotation matrix	
$\mathbf{W}$	Observation weighting matrix	
w	Weight function	
X	Precession-nutation coordinate X	
$\mathbf{x}$	State vector	
x	General variable	
dX	Correction to precession-nutation coordinate X, including FCN	
$\hat{\mathbf{x}}$	Best estimate of state vector	
x_p	Polar motion x coordinate	rad
Y	Precession-nutation coordinate Y	
$\mathbf{y}$	Observation vector	
y	General variable	
dY	Correction to precession-nutation coordinate Y, including FCN	
y_p	Polar motion y coordinate	rad
Z	Random variable	
$\mathbf{Z}$	Fisher information matrix	
z	Precession angle (rotation about inertial z-axis)	
z	General variable	
$\mathbf{z}$	Vector of random variables	
α	Fraction of stepsize	
α	View angle	
χ_0	Tuning parameter for Levenberg-Marquardt method	
∇	Backward difference operator	
Δ	Forward difference operator	
δ	Central difference operator	
δ_{0m}	Kronecker delta as a function of the geopotential order m	
ϵ	Tolerance in stepsize control	
ϵ_m	Machine Epsilon	
$\Delta\epsilon$	Nutation in the obliquity of the ecliptic	rad
ϵ	Observation error	
β	Simplification of step size function	

η	General variable	
f	Longitude function used in geopotential	
g	Derivative of longitude function used in geopotential	
Γ	Auxiliary stepsize function	
γ^*	Difference between constant stepsize coefficients	
γ	Constant stepsize coefficients	
κ	Earth albedo model switch: $\kappa = 1$ if illuminated, $\kappa = 0$ otherwise	
ξ	General variable	
λ	Störmer-Cowell predictor coefficient	
λ	Longitude	rad
λ_{LM}	Levenberg-Marquardt method damping parameter	
λ_M	Mean longitude	rad
μ	Specific gravitational constant (Earth, if without index)	km^3/s^2
μ	Mean value	
ν	Shadow function, discriminating between the satellite being in umbra, penumbra or full sunlight	
θ_{inc}	Solar incidence angle wrt. Earth plate surface normal	rad
$\mathbf{\Phi}$	State error transition matrix	
Φ	Solar flux	W/m^2
ϕ	Latitude	rad
ϕ	Absolute value of modified divided differences vector	m/s^2
$\boldsymbol{\phi}$	Modified divided differences vector	m/s^2
ϕ_{inc}	Solar incidence angle wrt. satellite surface normal	rad
Π	Transformation between normalized and unnormalized quantities of degree n and order m of the geopotential	
π	Mathematical constant ($\pi = 3.1415\ldots$)	
ψ	Angular distance of two points on unit sphere	rad
$\Delta\psi$	Nutation in longitude of the ecliptic	rad
ψ	Sum of steps	s
ρ	Density	kg/m^2
ρ	Radius function used in geopotential	km/s^2
ρ	Slant range	km
ρ	Fraction of next to current step size	
ρ_{LM}	Levenberg-Marquardt gain ratio	
σ	Standard deviation	
σ	Step size function	
τ	Time variable	s
τ	Time interval fraction	
θ_{ERA}	Earth rotation angle ERA	rad

θ_{GAST}	Earth rotation angle based on Greenwich apparent sidereal time	rad
Ω	Mean longitude of the ascending node of the Moon (Delaunay variable)	rad
Ω	Right ascension of ascending node	rad
ω	Argument of perigee	rad
$\omega_\oplus$	Angular velocity vector of Earth's rotation in inertial space, its magnitude being $\lvert\omega_\oplus\rvert = 7.292\,115\,0 \times 10^{-5}\,\mathrm{rad\,s^{-1}}$	rad/s
ζ	Precession angle (rotation about z-axis in MOD frame)	
ζ	Approximated error	
$3b$	Third body	
abs	Absorbed	
bf	Body-fixed	
C	Covariance	
$c1$	First shadow boundary crossing	
$c2$	Second shadow boundary crossing	
c	Compression	
c	Cross-section	
cfg	Configuration	
chs	Chaser	
cmb	Combined	
D	Atmospheric drag	
D	Mean elongation of the Moon from the Sun (Delaunay variable)	rad
d	Dimension	
d	Double integration	
DC	Differential Correction	
d_M	Mahalanobis distance	
e	End	
eph	Ephemeris	
f	File	
gc	Geocentric	
g	Geopotential	
h	Header	
I	Interpolation	
i	Counter	
id	Identifier/ID	
int	Integration	
JD	Julian Day	
l	Local	
λ	Longitude	

m	Message
man	Maneuver
max	Maximum
nam	Name
ns	Non-spherical
P	Plate
p	Perigee
p	Perturbation
p	Phase
p	Predicted value
ϕ	Geocentric latitude
pol	Polynomial
prj	Projection
r	Reference
r	Rings
r,d	Diffuse reflection
rel	Relative
r,s	Specular reflection
S	Satellite
S	State
s	Sample
s	Single integration
s	Start
seg	Segment
srf	Surface
tgt	Target
th	Threshold
U	Radial component
V	Along-track component
W	Cross-track or orbit normal component
w	Water
w	Wind
x	x-component of inertial reference frame
y	y-component of inertial reference frame
z	z-component of inertial reference frame
⊕	Earth
♃	Jupiter
♂	Mars
☿	Mercury
☾	Moon
♆	Neptune

♄	Saturn
☉	Sun
⛢	Uranus
♀	Venus

List of Abbreviations

List of Abbreviations

AC	Auto Configuration
ACT	Auto Configuration Table
AEL	Accepted Error Level
AEOS	Advanced Electro-Optical System
AFB	Air Force Base
AFSSS	Air Force Space Surveillance System
AGI	Analytical Graphics Inc.
API	Application Programming Interface
ARES	Assessment of Risk Event Statistics
ASCII	American Standard Code for Information Interchange
AU	Astronomical Unit, 1 $AU = 1.496 \cdot 10^{11}$ m
BOL	Begin of Life (of a spacecraft)
CA	Collision Avoidance
CCSDS	Consultative Committee for Space Data Systems
CDM	Conjunction Data Message
CHAMP	Challenging Minisatellite Payload
CIO	Celestial Intermediate Origin
CIP	Celestial Intermediate Pole
CIRF	Celestial Intermediate Reference Frame
CIRS	Celestial Intermediate Reference System
CLS	Collecte Localisation Satellites
cOEM	Encoded Orbit Ephemeris Message
cOMM	Encoded Orbit Mean-Element Message
ComSpOC	Commercial Space Operations Center
COSPAR	Committee on Space Research
CS	Cartesian Space
CSM	Conjunction Summary Message
CSR	Center for Space Research
CT	Correlated Track
DARPA	Defense Advanced Research Projects Agency

DC	Differential Correction
DE	Development Ephemerides
DE	JPL Development Ephemeris
DMC	Dynamic Model Compensation
DoD	Department of Defense
DOF	Degrees-of-Freedom
DORIS	Doppler Orbitography and Radiopositioning Integrated by Satellite
Dst	Disturbance storm time
ECI	Earth Centered Inertial
EGM	Earth Gravitational Model
EIGEN	European Improved Gravity model of the Earth by New techniques
EISCAT	European Incoherent Scatter Scientific Association
EKF	Extended Kalman Filter
EOP	Earth Orientation Parameters
ERA	Earth Rotation Angle
ERP	Earth Radiation Pressure
ERS	European Remote Sensing satellite
ESA	European Space Agency
ESOC	European Space Operations Centre
EU	European Union
EWS	Early Warning System
FCN	Free Core Nutation
FK	Fundamental Katalog
FTP	File Transfer Protocol
GAMBIT	Geopotential Adaptation Method to Bias Trajectories
GAST	Greenwich Apparent Sidereal Time
GCRF	Geocentric Celestial Reference Frame
GCRS	Geocentric Celestial Reference System
GEO	Geostationary Earth Orbit
GEODSS	Ground-based Electro-Optical Deep-Space Surveillance
GFZ	Deutsches GeoForschungsZentrum
GLONASS	Globalnaya navigatsionnaya sputnikovaya sistema
GOCE	Gravity Field and Steady-State Ocean Circulation Explorer
GP	General Perturbations
GPS	Global Positioning System
GRACE	Gravity Recovery and Climate Experiment

GRACE-FO	GRACE Follow-on
GRAVES	Grand Réseau Adapté a la Veille Spatiale
GTO	Geostationary Transfer Orbit
HEO	High Eccentricity Orbit
HP	High Precision
HPOP	High Precision Orbit Propagator
HWM	Horizontal Wind Model
IAS	Institute of Aerospace Systems
IAU	International Astronomical Union
ICESat	Ice, Cloud, and Land Elevation Satellite
ICGEM	International Centre for Global Earth Models
ICRF	International Celestial Reference Frame
ICRS	International Celestial Reference System
IDS	International DORIS Service
IERS	International Earth Rotation and Reference Systems Service
IOD	Initial Orbit Determination
IR	Infrared
IRM	IERS Reference Meridian
IRP	IERS Reference Pole
ISO	International Organization for Standardization
ISON	International Scientific Optical Network
ISS	International Space Station
ITRF	International Terrestrial Reference Frame
ITRS	International Terrestrial Reference System
IUGG	International Union of Geodesy and Geophysics
JAXA	Japan Aerospace Exploration Agency
JPL	Jet Propulsion Laboratory
JSpOC	Joint Space Operations Center
LAGEOS	Laser Geodynamics Satellites
LEGOS	Laboratoire d'Etudes en Géophysique et Océanographie Spatiales
LEO	Low Earth Orbit
LEOP	Launch and Early Orbit Phase
LM	Levenberg-Marquardt method
LOD	Length of Day
LSB	Least Significant Bit

MASTER	Meteoroid and Space Debris Terrestrial Environment Reference
MC	Monte Carlo
MEO	Medium Earth Orbit
MJD	Modified Julian Day
MOD	Mean of Date
MOTIF	Maui Optical Tracking and Identification Facility
MSB	Most Significant Bit
MSIS	Mass Spectrometer and Incoherent Scatter Radar
MSSS	Maui Space Surveillance Site
NAIF	Navigation and Ancillary Information Facility
NANU	Notice Advisory to NAVSTAR Users
NASA	National Aeronautics and Space Administration
NAVSTAR	Navigation System using Timing and Ranging
NDM	Navigation Data Message
NEPTUNE	Networking/Partnering Initiative (ESA research programme with industry and academia) (NPI) Ephemeris Propagation Tool with Uncertainty Extrapolation
NGA	National Geospatial-Intelligence Agency
NORAD	North American Aerospace Defense Command
NPI	Networking/Partnering Initiative (ESA research programme with industry and academia)
NRL	Naval Research Laboratory
NRLMSISE-00	Naval Research Laboratory Mass Spectrometer and Incoherent Scatter (Extended), Earth atmosphere model
NSO	Non-Singular Orbital Elements
NSSCC	National Space Surveillance Control Center
O/O	Owner and/or Operator
OCM	Orbital Conjunction Message
OCM	Orbit Comprehensive Message
OCRF	Orbit Centered Reference Frame
OD	Orbit Determination
ODE	Ordinary Differential Equation
ODIN	Orbit Determination by Improved Normal Equations
ODM	Orbit Data Message
ODR	Orbital Data Request
OEM	Orbit Ephemeris Message
OMM	Orbit Mean-Element Message
OPM	Orbit Parameter Message

PDF	Probability Density Function
PEC	Predict-Evaluate-Correct
PEF	Pseudo Earth Fixed
POD	Precise Orbit Determination
POE	Precision Orbit Ephemeris
PRARE	Precise Range And Range-rate Equipment
PRN	Pseudo Random Noise
RAAN	Right Ascension of Ascending Node
RCS	Radar Cross-Section
RK	Runge-Kutta
RMS	Root Mean Square
RSS	Residuals Sum of Squares
RTN	Radial, Transversal, Normal
SA	Selective Availability
SBSS	Space Based Space Surveillance
SBTC	Shadow Boundary Transit Correction
SDP	Simplified Deep-space Perturbations
SFU	Solar Flux Units (1 sfu = 1.0×10^2)
SGP	Simplified General Perturbations
SLR	Satellite Laser Ranging
SNC	State Noise Compensation
SOFA	Standards Of Fundamental Astronomy
SP	Special Perturbations
SP3	Standard Product # 3
SPA	Support to Precursor Space Situational Awareness Services
SPICE	Spacecraft Planet Instrument C-matrix Events
SPK	SPICE Kernel
SQL	Structured Query Language
SRP	Solar Radiation Pressure
SSA	Space Situational Awareness
SSA-PP	SSA Preparatory Programme
SSN	Space Surveillance Network
SST	Space Surveillance Telescope
SST	Space Surveillance and Tracking
STS	Space Transportation System
SV	State Vector
SVN	Subversion
SVN	Satellite Vehicle Number

TAI	International Atomic Time (*Temps Atomique International*)
TCA	Time of Closest Approach
TDB	Barycentric Dynamical Time (*Temps Dynamique Barycentric*)
TDT	Terrestrial Dynamical Time
TEME	True Equator Mean Equinox
TGP	Tide Generating Potential
TIO	Terrestrial Intermediate Origin
TIP	Tracking and Impact Prediction
TIRA	Tracking and Imaging Radar
TIRF	Terrestrial Intermediate Reference Frame
TIRS	Terrestrial Intermediate Reference System
TLE	Two Line Elements
TOD	True of Date
TT	Terrestrial Time
TU-BS	Technische Universität Braunschweig
UCT	Uncorrelated Track
UKF	Unscented Kalman Filter
US	United States
USG	United States Government
USSTRATCOM	United States Strategic Command
UT	Universal Time
UT1	Universal Time corrected for polar motion
UTC	Universal Time Coordinated
VLBI	Very Long Baseline Interferometry
WGS	World Geodetic System
XML	Extensible Markup Language
XSD	XML Schema Definition
ZUNIEM	Zuschlag Numerical Integration of the Equations of Motion

A

State-of-the-art propagation with the NEPTUNE software

This annex describes the models and methods that were implemented in the numerical integration tool NEPTUNE based on current standards and best practices for orbit propagation, which were:

- ISO 8601:2004, Data elements and interchange formats – Information interchange – Representation of dates and times (ISO, 2004),
- AIAA G-003C-2010, Guide to Reference and Standard Atmosphere Models (AIAA, 2010),
- Astrodynamics - Propagation Specifications, Technical Definitions, and Recommended Practices (ANSI/AIAA, 2010) and
- the European standard on the space environment, ECSS-ST-10-04C (ECSS, 2008), has been developed by the European Space Agency, national space agencies and European industry associations.

A.1. Units, precision, time, constants and coordinates

A.1.1. Units

SI units are used throughout the software, as well as units that are approved for use with SI, like angular degrees (AIAA, 2010).

A.1.2. Precision

Constants, such as π, are derived with machine precision and used to compute other constants, like the conversion ratio from angles in degrees to radian (AIAA, 2010).

Computations with floating point numbers are performed with double precision, where possible, to preserve numerical precision (AIAA, 2010).

A.1.3. Time

For each output, Universal Time Coordinated (UTC) time tags according to ISO 8601:2004 (ISO, 2004) are provided, including leap seconds.

The difference between UT1 and UTC ($\Delta UT1$), as well as the excess length of day (LOD) are obtained via the EOP obtained from IERS (ANSI/AIAA, 2010).

A.1.4. Constants

Constants, like μ are applied consistent with the models they have been derived from. For example, the Earth's equatorial radius and the specific gravitational constant, μ, are adopted from the geopotential model (ANSI/AIAA, 2010).

Astronomical constants are taken from the Astronomical Almanac.

A.1.5. Coordinates

The GCRF is used for the satellite orbit integration, while accelerations in an Earth-fixed frame are computed with respect to the ITRF. The conversion between these two frames is based on the CIO-approach, using the complete reduction in every time step (ANSI/AIAA, 2010). For increased performance, the precession-nutation parameters X, Y and s are interpolated with a polynomial interpolation of 5th degree (ANSI/AIAA, 2010).

A.2. Integration

The Störmer predictor and Cowell corrector method (PEC) is used, with variable step numerical integration, which is beneficial for high-eccentricity orbits. The integration is also a self-starting multi-step and double-integration, which means to compute position directly from the acceleration.

A.3. Force model

A.3.1. Geopotential

The European model EIGEN-GL04C (ECSS, 2008), as well as the models EGM96 and EGM2008 are implemented in Neptune , with the latter two being recommended by (ANSI/AIAA, 2010).

A.3.2. Atmospheric drag

The last 81-day average of the solar activity index $F_{10.7}$ is used with the NRLMSISE-00 model, which is recommended by ECSS, 2008. However, it is used for all altitudes, while ECSS, 2008 recommends to use JB-2006 for altitudes above 120 km. The HWM07 wind model is used, where (ECSS, 2008) recommends the -93 version. The input for geomagnetic activity are the A_p indices, which provide additional sensitivity not available in the K_p indices (ANSI/AIAA, 2010). The three-hourly A_p's are interpolated using cubic splines, a method discussed by Vallado and Kelso (2005) (ANSI/AIAA, 2010).

Using the daily $F_{10.7}$ indices is with respect to the time the measurement was actually taken. The offset (17:00 UTC until 1991-05-31, 20:00 UTC afterwards) is considered (ANSI/AIAA, 2010).

A.3.3. Third-body gravitation

The DE405 (ECSS, 2008) and DE421 series are used to obtain the ephemerides for the solar system bodies, which are consistent with the IAU 2000 theory (ANSI/AIAA, 2010).

A.3.4. Solar radiation pressure

A conical shadow model including Umbra and Penumbra (dual-cone) (ISO, 2014) for a spherical Earth is used. It is possible to define a macro model and simple attitude motion laws in Neptune . The solar luminosity is computed as a function of the distance between Sun and Earth, which is a seasonal variation. The numerical integrations accounts for a correction of shadow boundary crossings, by determining the exact times of Penumbra and Umbra exit and entry, respectively.

A.3.5. Earth radiation pressure

In Neptune , also the Earth albedo, the reflected sunlight in the optical wavelengths, and the Earth emissivity in the infrared are considered. The model by Knocke (1989) is used (as recommended by (ECSS, 2008)), with modeling the Earth as seen from the satellite by a set of plates with individual viewing angles and latitude- and time-dependent albedo.

A.3.6. Tides

The models for Solid Earth and pole tides have been implemented in Neptune according to the IERS conventions (Luzum and Petit, 2010) (ECSS, 2008). Solid Earth tidal contributions are modelled as time-varying spherical harmonic coefficients. Care has been taken to include only tide-free geopotential models. For future updates, this has to be kept in mind, as sometimes, the J_2 term already contains the permanent tide part (ANSI/AIAA, 2010).

A very simple ocean tide model has been used in Neptune so far according to Vallado and McClain (2013). However, it would be worth considering adding more recent and modern models, examples being the CSR4 (Eanes and Bettadpur, 1995) or the FES2004 (Letellier et al., 2004) models.

A.3.7. Other

The propagator Neptune has been developed with its application in the SST segment in mind. Therefore, very accurate orbit determination, which would involve active means like GPS, SLR or DORIS, are for most objects not available. Also, object properties like size, mass, shape and aerodynamic coefficients are, in general, unknown. Hence, perturbations which contribute in the sub-metre regime have been disregarded for the time being. However, if Neptune would be considered also for high-precision orbit determination, one can always think of adding additional modules to account for perturbations like the General Relativity, thermal forces (Yarkovsky effects) or even antenna thrust.

In its current implementation, Neptune allows to include manoeuvres into the computation. They are modelled as additional accelerations, which are added to an input file and incorporated into the integration during runtime.

B

Numerical integration

This chapter covers in more detail the theoretical background and especially the equations behind the numerical integration in NEPTUNE to support and complement the algorithm description in the main part of the document.

B.1. State vector integration using Berry's Störmer-Cowell method

The full derivation of the variable-step double-integration multi-step Störmer-Cowell integration is given by Berry (2004). Only parts thereof are provided here for convenience. The interested reader is highly recommended to study Berry (2004) for a full theoretical treatise of this subject.

The differential equation to be solved (Equation 2.1) is:

$$\ddot{\mathbf{r}} = -\frac{\mu}{r^2}\frac{\mathbf{r}}{r} + \mathbf{a}_p = \mathbf{f}(t, \mathbf{r}, \dot{\mathbf{r}}). \tag{B.1}$$

For a multi-step integration, the k function values $\mathbf{f}_{n-k+1} \dots \mathbf{f}_n$ are interpolated by a $(k-1)^{th}$ degree polynomial:

$$\mathbf{P}_{k,n}(t_{n+1-i}) = \mathbf{f}_{n+1-i}, \quad i = 1 \dots k. \tag{B.2}$$

B.1.1. Divided differences

In numerical integration, finite differences are typically used to solve differential equations when the function derivatives are not available. There are three different forms for finite differences with an integration stepsize h:

1. Forward difference:

$$\Delta f(x) = f(x+h) - f(x) \tag{B.3}$$

2. Central difference:

$$\delta f(x) = f\left(x + \frac{1}{2}h\right) - f\left(x - \frac{1}{2}h\right) \tag{B.4}$$

3. Backward difference:

$$\nabla f(x) = f(x) - f(x-h) \tag{B.5}$$

As an example, consider the Störmer predictor expressed in terms of backward differences:

$$\mathbf{r}_{n+1} = 2\mathbf{r}_n + \mathbf{r}_{n-1} + h^2 \left(1 + \frac{1}{12}\nabla^2 + \frac{1}{12}\nabla^3 + \frac{19}{240}\nabla^4 + \frac{3}{40}\nabla^5 + \cdots\right)\ddot{\mathbf{r}}_n \tag{B.6}$$

The above predictor is double-integration, as the position vector $\mathbf{r}_{n+1}$ is obtained directly from the acceleration $\ddot{\mathbf{r}}$ of the previous step. However, it is for a constant stepsize h. For a multistep integrator therefore a restart is required each time the stepsize changes. This is unfavourable, especially if the evaluation of the force function is costly. A solution was proposed by Krogh (1974) using *divided differences*. The advantage is that with divided differences, the coefficients used in the integration need to be re-computed each time the stepsize changes, which is by far more efficient than re-computing finite differences, if the function evaluations are costly.

The divided differences are calculated through a recursive relation (Berry, 2004):

$$f[t_n] = f_n, \tag{B.7}$$

$$f[t_n, t_{n-1}] = \frac{f_n - f_{n-1}}{t_n - t_{n-1}}, \tag{B.8}$$

$$f[t_n, \ldots, t_{n-i}] = \frac{f[t_n, \ldots, t_{n-i+1}] - f[t_{n-1}, \ldots, t_{n-i}]}{t_n - t_{n-i}}. \tag{B.9}$$

The interpolating polynomial in Equation B.2 can be written in divided difference form:

$$\begin{aligned}\mathbf{P}_{k,n}(t) &= f[t] + (t - t_n) f[t, t_{n-1}] + \ldots + \\ &+ (t - t_n)(t - t_{n-1}) \ldots (t - t_{n-k+2}) f[t, t_{n-1}, \ldots, t_{n-k+1}].\end{aligned} \tag{B.10}$$

The *modified divided differences* are introduced as:

$$\boldsymbol{\phi}_1(n) = \mathbf{f}[t_n] = \ddot{r}_n \tag{B.11}$$

$$\boldsymbol{\phi}_i(n) = \psi_1(n)\psi_2(n) \ldots \psi_{i-1}(n)\mathbf{f}[t_n, t_{n-1}, \ldots, t_{n-i+1}] \quad i > 1, \tag{B.12}$$

where

$$\psi_i(n) = h_n + h_{n-1} + h_{n-2} + \cdots + h_{n+1-i}. \tag{B.13}$$

Berry (2004) further introduces:

$$\boldsymbol{\phi}_i^*(n) = \beta_i(n+1)\boldsymbol{\phi}_i(n), \tag{B.14}$$

where

$$\beta_1(n+1) = 1, \tag{B.15}$$

$$\beta_i(n+1) = \frac{\psi_1(n+1)\,\psi_2(n+1) \cdots \psi_{i-1}(n+1)}{\psi_1(n)\,\psi_2(n) \ldots \psi_{i-1}(n)} \quad i > 1. \tag{B.16}$$

B.1.2. Predictor

For a single-integration the velocity is obtained from the integration of the interpolating polynomial, which is the predictor of the Shampine-Gordon integrator (Shampine and Gordon, 1975):

$$\dot{\mathbf{r}}^{p}_{n+1} = \dot{\mathbf{r}}_n + \int_{t_n}^{t_{n+1}} \mathbf{P}_{k,n}(t)\,dt \tag{B.17}$$

The double-integration method used for the Störmer predictor is used to obtain the radius vector directly from the accelerations (while the velocity is obtained via the Shampine-Gordon predictor):

$$\mathbf{r}^{p}_{n+1} = \mathbf{r}_n + h_{n+1}\dot{\mathbf{r}}_n + \int_{t_n}^{t_{n+1}} \int_{t_n}^{\bar{t}} \mathbf{P}(t)\,d\bar{t}dt. \tag{B.18}$$

Now the interpolating polynomial can be replaced by a series representation using Equation B.14:

$$\mathbf{P}_{k,n}(t) = \sum_{i=1}^{k} c_{i,n}(\tau)\,\boldsymbol{\phi}^{*}_{i}(n), \tag{B.19}$$

where τ is the fraction of the current interval:

$$\tau = \frac{t - t_n}{h_{n+1}}. \tag{B.20}$$

After performing the integration of Equation B.18 using Equation B.19 one obtains:

$$\mathbf{r}^{p}_{n+1} = \left(1 + \frac{h_{n+1}}{h_n}\right)\mathbf{r}_n - \frac{h_{n+1}}{h_n}\mathbf{r}_{n-1} + h^2_{n+1}\sum_{i=1}^{k}\left(g_{i,2} + \frac{h_{n+1}}{h_n}g'_{i,2}\right)\boldsymbol{\phi}^{*}_{i}(n), \tag{B.21}$$

with the coefficients $g'_{i,2}$, which are dependent on the stepsizes for the backpoints, computed via:

$$g'_{i,q} = \begin{cases} \frac{1}{q}\left(\frac{-h_n}{h_{n+1}}\right)^{q} & i = 1, \\ \frac{1}{q(q+1)}\left(\frac{-h_n}{h_{n+1}}\right)^{q+1} & i = 2, \\ \frac{\psi_{i-3}(n-1)}{\psi_{i-1}(n+1)}g'_{i-1,q} - \alpha_{i-1}(n+1)\,g'_{i-1,q+1} & i > 2, \end{cases} \tag{B.22}$$

and the coefficients $g_{i,q}$ (which are constant, also for variable-step integration) via:

$$g_{i,q} = \begin{cases} \frac{1}{q} & i = 1, \\ \frac{1}{q(q+1)} & i = 2, \\ g'_{i-1,q} - \alpha_{i-1}(n+1)\,g'_{i-1,q+1} & i > 2. \end{cases} \tag{B.23}$$

The auxiliary variable α is a fraction of the stepsizes:

$$\alpha_i(n+1) = \frac{h+1}{\psi_i(n+1)}. \tag{B.24}$$

A second integration is required to obtain the velocity vector. The Shampine-Gordon predictor formulation is obtained using the Equation B.17:

$$\dot{\mathbf{r}}^{p}_{n+1} = \dot{\mathbf{r}}_n + h_{n+1} \sum_{i=1}^{k} g_{i,1} \boldsymbol{\phi}^{*}_{i}(n). \tag{B.25}$$

B.1.3. Corrector

After a predicted value $\mathbf{r}^{p}_{n+1}$ is found, a new function evaluation at that point provides a predicted acceleration. The interpolation polynomial can then be one degree higher than $\mathbf{P}_{k,n}(t)$ and the new formulation using the series representation from Equation B.19 is:

$$\mathbf{P}^{*}_{k+1,n}(t) = \mathbf{P}_{k,n}(t) + c_{k+1,n}(\tau)\, \boldsymbol{\phi}^{p}_{k+1}(n+1). \tag{B.26}$$

Using the same approach as for the predictor, the integration results in the variable-step Cowell predictor:

$$\mathbf{r}_{n+1} = \mathbf{r}^{p}_{n+1} + h^{2}_{n+1} \left(g_{k+1,2} + \frac{h_{n+1}}{h_n} g'_{k+1,2} \right) \boldsymbol{\phi}^{p}_{k+1}(n+1). \tag{B.27}$$

Similarly, the velocity vector is obtained from the Shampine-Gordon corrector by replacing the polynomial in Equation B.17 with Equation B.26, which ultimately results in:

$$\dot{\mathbf{r}}_{n+1} = \dot{\mathbf{r}}^{p}_{n+1} + h_{n+1} g_{k+1,1} \boldsymbol{\phi}^{p}_{k+1}(n+1). \tag{B.28}$$

B.1.4. Stepsize control

For the variable-step integration, the stepsize is controlled at each step keeping the local error below a user-defined tolerance, which is defined as a combination of an absolute tolerance, ϵ_{abs}, and a relative tolerance, ϵ_{rel}, for the individual position and velocity components:

$$\epsilon^{s}_{l,i} \leqslant \epsilon_{rel} \dot{r}_i + \epsilon_{abs}, \tag{B.29}$$

$$\epsilon^{d}_{l,i} \leqslant \epsilon_{rel} r_i + \epsilon_{abs}, \tag{B.30}$$

where the index s is for single integration (velocity from acceleration) and d is for double integration (position from acceleration), respectively. Note that this formulation guarantees that there is a lower boundary on the user-defined error provided by ϵ_{abs}.

In order to compute the local error at each step, the corrector result from Equation B.27 (k^{th} degree polynomial) is compared with a corrector using the set of $n+1$ points interpolated by a polynomial of degree k:

$$\epsilon^{s}_{l} = \dot{\mathbf{r}}_{n+1} - \dot{\mathbf{r}}_{n+1}(k), \tag{B.31}$$

$$\epsilon^{d}_{l} = \mathbf{r}_{n+1} - \mathbf{r}_{n+1}(k). \tag{B.32}$$

Using the $(k-1)^{th}$ degree polynomial from Equation B.19, that now passes through the additional point resulting from the corrector, performing the same integration and computing the difference for the local error in Equation B.31, one obtains (for the single integration):

$$\epsilon_l^s(k) \approx h_{n+1}(g_{k+1,1} - g_{k,1})\,\boldsymbol{\phi}_{k+1}^p(n+1). \tag{B.33}$$

The components of the local error are combined in a weighted sum of squares at each step and compared with the tolerance ϵ_{max}:

$$\sqrt{\sum_{i=1}^{3}\left(\frac{\epsilon_{l,i}^s}{w_i^s}\right)^2} \leqslant \epsilon_{max}, \tag{B.34}$$

with

$$\epsilon_{max} = \max(\epsilon_{rel}, \epsilon_{abs}), \tag{B.35}$$

and the weight functions

$$w_i^s = |\dot{r}_i|\frac{\epsilon_{rel}}{\epsilon_{max}} + \frac{\epsilon_{abs}}{\epsilon_{max}}. \tag{B.36}$$

The step is successful if Equation B.34 is fulfilled, otherwise the step is repeated with half the stepsize. If this happens three times consecutively, the integration is reset and starts as a first-order method again. In general, such a restart will be required at discontinuities like shadow boundary crossings, manoeuvres, etc.

After a step was successful, the stepsize for the next step is computed to keep the local error as close as possible to the tolerance. Using $h_{n+2} = \rho h_{n+1}$, assuming that the divided differences are slowly varying and that all preceding steps were taken with h_{n+2}, Berry (2004) gives the local error for the following step as:

$$\epsilon_{l,n+2}^s(k) = \rho^{k+1} h_{n+1}\gamma_k^*\sigma_{k+1}(n+1)\,\boldsymbol{\phi}_{k+1}^p(n+1), \tag{B.37}$$

with γ^* being the difference between the constant step size coefficients

$$\gamma^* = \gamma_k - \gamma_{k-1}, \tag{B.38}$$

and the auxiliary functions $\sigma_{k+1}(n+1)$ being found via a recursion:

$$\sigma_1(n+1) = 1, \tag{B.39}$$

$$\sigma_i(n+1) = (i-1)\,\alpha_{i-1}(n+1)\,\sigma_{i-1}(n+1), \quad i > 1, \tag{B.40}$$

where α_i is the fraction of the current step size to the sum of the current step and the previous ones:

$$\alpha_i(n+1) = \frac{h_{n+1}}{h_{n+1} + h_n + \ldots + h_{n+2-i}}. \tag{B.41}$$

The approximated local error ζ_l^s that would be made if the previous steps had been taken with h_{n+1} results as:

$$\zeta_l^s = |h_{n+1}\gamma_k^*\sigma_{k+1}(n+1)|\sqrt{\sum_{i=1}^{3}\left(\frac{\phi_{i,k+1}^p(n+1)}{w_{l,i}^s}\right)^2}. \tag{B.42}$$

The stepsize factor ρ can then be computed from

$$\rho^s = \left(\frac{\epsilon_{max}}{\zeta_l^s}\right)^{\frac{1}{k+1}} \tag{B.43}$$

In order to compensate for the assumptions in the derivation of the stepsize factor, Shampine and Gordon (1975) introduced a so-called *chicken factor*, resulting in a more conservative estimate:

$$\rho^s = \left(\frac{0.5\epsilon_{max}}{\zeta_l^s}\right)^{\frac{1}{k+1}} \tag{B.44}$$

The calculated value of ρ is bounded between 0.5 and 2.0, so that the stepsize is doubled for all values $\rho \geqslant 2.0$ and halved for $\rho \leqslant 0.5$.

While the above derivation is for single integration, Berry (2004) provides the formulae for the double integration analogously. The local error vector for double integration can be computed from:

$$\epsilon_l^d(k) \approx h_{n+1}^2 \left(g_{k+1,2} - g_{k,2} + \frac{h_{n+1}}{h_n}\left((g'_{k+1,2} - g'_{k,2}\right)\right) \phi_{k+1}^p(n+1). \tag{B.45}$$

The local error components are weighted and combined for the comparison against the tolerance:

$$\sqrt{\sum_{i=1}^{3}\left(\frac{\epsilon_{l,i}^d}{w_i^d}\right)^2} \leqslant \epsilon_{max}, \tag{B.46}$$

with the weights for double integration obtained from

$$w_i^d = |r_i|\frac{\epsilon_{rel}}{\epsilon_{max}} + \frac{\epsilon_{abs}}{\epsilon_{max}}. \tag{B.47}$$

Using both, single integration to obtain the velocity vector and double integration for the position vector, a step fails if either the combined weighted local error components from Equation B.34 or Equation B.46 are below the tolerance (Equation B.35).

Similarly, Berry (2004) gives the estimated local error for the next step in order to derive the stepsize factor ρ:

$$\zeta_l^d = \left|h_{n+1}^2 \lambda_k^* \sigma_{k+1}(n+1)\right| \sqrt{\sum_{i=1}^{3}\left(\frac{\phi_{i,k+1}^p(n+1)}{w_{l,i}^d}\right)^2}, \tag{B.48}$$

where λ_k^* are the coefficients of the Störmer-Cowell predictor coefficients. With the same chicken factor of 0.5, the stepsize factor for double integration can be computed as:

$$\rho^d = \left(\frac{0.5\epsilon_{max}}{\zeta_l^d}\right)^{\frac{1}{k+1}} \tag{B.49}$$

The stepsize factor actually used in the integration is the minimum of the values obtained for single and double integration:

$$\rho = \min\left(\rho^s, \rho^d\right). \tag{B.50}$$

While Shampine and Gordon (1975) designed their method in a way preferring constant stepsizes, thereby reducing the overhead for the re-computation of the integrator coefficients, Berry (2004) correctly points out that due to the very expensive force model, the additional integrator overhead can be neglected in favour of having fast increase towards larger stepsizes. Therefore, Berry (2004) recommended to have the same boundaries at $\rho = 0.5$ and $\rho = 2.0$, while not having any other restrictions for values in between.

B.1.5. Interpolation

In principle, the integrator can be configured to have a stepsize corresponding to the requested output time. Meaning that the stepsize would then be smaller than possible for a given error tolerance, this would introduce unnecessary computation cost. Therefore, Shampine and Gordon (1975) give an interpolation formula to find the output value at the requested time t_I, which is between the points n and $n+1$. Using the $(k+1)^{th}$ degree polynomial for the set of backpoints, the interpolated value is found via (Shampine and Gordon, 1975):

$$\dot{\mathbf{r}}_I = \dot{\mathbf{r}}_{n+1} + \int_{t_{n+1}}^{t_I} \mathbf{P}_{k+1,n+1}(t)\, dt. \tag{B.51}$$

Performing a similar derivation as for the predictor (Section B.1.2), one obtains for the single integration:

$$\dot{\mathbf{r}}_I = \dot{\mathbf{r}}_{n+1} + h_I \sum_{i=1}^{k+1} g^I_{i,1} \boldsymbol{\phi}_i (n+1), \tag{B.52}$$

with the interpolation stepsize $h_I = t_I - t_{n+1}$ and the coefficients $g^I_{i,1}$ which can be computed via a recursion:

$$g^I_{i,q} = \begin{cases} \dfrac{1}{q} & i = 1, \\ \Gamma_{i-1}(1)\, g^I_{i-1,q} - \frac{h_I}{\psi_{i-1}(n+1)} g^I_{i-1,q+1} & i \geqslant 2. \end{cases} \tag{B.53}$$

The auxiliary stepsize functions Γ_{i-1} are computed with the stepsize fraction τ as the independent variable:

$$\Gamma_i(\tau) = \begin{cases} \dfrac{\tau h_I}{\psi_1(n_1)} & i = 1, \\ \frac{\tau h_I + \psi_{i-1}(n+1)}{\psi_i(n+1)} & i \geqslant 2, \end{cases} \tag{B.54}$$

where

$$\tau = \frac{t - t_{n+1}}{h_I}. \tag{B.55}$$

Analogously, Berry (2004) derives the interpolation formulae for the double integration:

$$\mathbf{r}_I = \left(1 + \frac{h_I}{h_{n+1}}\right) \mathbf{r}_{n+1} - \frac{h_I}{h_{n+1}} \mathbf{r}_n + h_I^2 \sum_{i=1}^{k+1} \left(g^I_{i,2} + \frac{h_I}{h_{n+1}} g^{I'}_{i,2} \right) \boldsymbol{\phi}_i(n+1), \tag{B.56}$$

where

$$g^{I'}_{i,q} = \begin{cases} \dfrac{1}{q} \left(\dfrac{-h_{n+1}}{h_I} \right) & i = 1, \\ \Gamma_{i-1}\left(\frac{-h_{n+1}}{h_I}\right) g^{I'}_{i-1,q} - \frac{h_I}{\psi_{i-1}(n+1)} g^{I'}_{i-1,q+1} & i \geqslant 2. \end{cases} \tag{B.57}$$

C

Conversion between GCRF and ITRF

C.1. Precession and nutation (IAU 2006/2000)

The matrix $\mathbf{Q}(t)$ converts a state from the ICRF to the GCRF:

$$\mathbf{r}_{GCRF} = \mathbf{Q}(t)\,\mathbf{r}_{ICRF}. \tag{C.1}$$

It accounts for the motion of the CIP in the GCRF due to precession and nutation, which can be expressed as (Luzum and Petit, 2010):

$$\mathbf{Q}(t) = \begin{pmatrix} 1-aX^2 & -aXY & X \\ -aXY & 1-aY^2 & Y \\ -X & -Y & 1-a(X^2+Y^2) \end{pmatrix} \mathbf{T}_3(s), \tag{C.2}$$

with $\mathbf{T}_3(s)$ being a rotation providing the position of the CIO on the equator of the CIP and the angle s is therefore called the "CIO locator" (Luzum and Petit, 2010). The quantity a is computed as (Luzum and Petit, 2010):

$$a = \frac{1}{1+\sqrt{1-X^2-Y^2}} \simeq \frac{1}{2} + \frac{1}{8}\left(X^2+Y^2\right). \tag{C.3}$$

The series for X, Y and s are given in the following, each containing a polynomial and a trigonometric part. Note also, that the CIO locator is a function of X and Y. The coefficients of the individual series terms are subject to future revisions and can be accessed via the website of the US Naval Observatory[1].

$$\begin{aligned} X =& -0.016\,617'' + 2004.191\,898''t - 0.429\,782\,9''t^2 - 0.198\,618\,34''t^3 + \\ &+ 0.000\,007\,578''t^4 + 0.000\,005\,928\,5''t^5 + \\ &+ \sum_{i=1}^{1306}\left[A_{xs0,i}\sin\left(a_{p,i}\right) + A_{xc0,i}\cos\left(a_{p,i}\right)\right] + \sum_{i=1}^{253}\left[A_{xs1,i}\sin\left(a_{p,i}\right) + A_{xc1,i}\cos\left(a_{p,i}\right)\right]t + \\ &+ \sum_{i=1}^{36}\left[A_{xs2,i}\sin\left(a_{p,i}\right) + A_{xc2,i}\cos\left(a_{p,i}\right)\right]t^2 + \sum_{i=1}^{4}\left[A_{xs3,i}\sin\left(a_{p,i}\right) + A_{xc3,i}\cos\left(a_{p,i}\right)\right]t^3 + \end{aligned} \tag{C.4}$$

[1] U.S. Naval Observatory (Table 5.2*a*, 5.2*b* and 5.2*d*) http://maia.usno.navy.mil/conv2010/conv2010_c5.html, last access on January 29, 2015.

$$+\sum_{i=1}^{1}\left[A_{xs4,i}\sin\left(a_{p,i}\right)+A_{xc4,i}\cos\left(a_{p,i}\right)\right]t^4$$

$$\begin{aligned}
Y = & -0.006\,951'' - 0.025\,896''t - 22.407\,274\,7''t^2 + 0.001\,900\,59''t^3 + \\
& + 0.001\,112\,526''t^4 + 0.000\,000\,135\,8''t^5 + \\
& + \sum_{i=1}^{962}\left[A_{ys0,i}\sin\left(a_{p,i}\right)+A_{yc0,i}\cos\left(a_{p,i}\right)\right] + \sum_{i=1}^{277}\left[A_{ys1,i}\sin\left(a_{p,i}\right)+A_{yc1,i}\cos\left(a_{p,i}\right)\right]t + \\
& + \sum_{i=1}^{30}\left[A_{ys2,i}\sin\left(a_{p,i}\right)+A_{yc2,i}\cos\left(a_{p,i}\right)\right]t^2 + \sum_{i=1}^{5}\left[A_{ys3,i}\sin\left(a_{p,i}\right)+A_{yc3,i}\cos\left(a_{p,i}\right)\right]t^3 + \\
& + \sum_{i=1}^{1}\left[A_{ys4,i}\sin\left(a_{p,i}\right)+A_{yc4,i}\cos\left(a_{p,i}\right)\right]t^4
\end{aligned} \tag{C.5}$$

$$\begin{aligned}
s = & -\frac{XY}{2} + 0.000\,094'' + 0.003\,808\,65''t - 0.000\,122\,68''t^2 - 0.072\,574\,11''t^3 + \\
& + 0.000\,027\,98''t^4 + 0.000\,015\,62''t^5 + \\
& + \sum_{i=1}^{33}\left[A_{ss0,i}\sin\left(a_{p,i}\right)+A_{sc0,i}\cos\left(a_{p,i}\right)\right] + \sum_{i=1}^{3}\left[A_{ss1,i}\sin\left(a_{p,i}\right)+A_{sc1,i}\cos\left(a_{p,i}\right)\right]t + \\
& + \sum_{i=1}^{25}\left[A_{ss2,i}\sin\left(a_{p,i}\right)+A_{sc2,i}\cos\left(a_{p,i}\right)\right]t^2 + \sum_{i=1}^{4}\left[A_{ss3,i}\sin\left(a_{p,i}\right)+A_{sc3,i}\cos\left(a_{p,i}\right)\right]t^3 + \\
& + \sum_{i=1}^{1}\left[A_{ss4,i}\sin\left(a_{p,i}\right)+A_{sc4,i}\cos\left(a_{p,i}\right)\right]t^4
\end{aligned} \tag{C.6}$$

The arguments $a_{p,i}$ in the trigonometric series are a linear combination of 14 different terms to account for luni-solar and planetary nutation. The time variable in the following is also TT, while the original equations are based on Barycentric Dynamical Time (*Temps Dynamique Barycentric*) (TDB), the latter being defined as "the independent argument of ephemerides and equations and motion that are referred to the barycenter of the solar system" (Seidelmann, 2006). Using TT instead of TDB, however, results in CIP location errors less than 0.01 μas (Luzum and Petit, 2010), which are negligible. The full equation, as provided by Vallado and McClain (2013), using the Delaunay variables for the Sun and the Moon:

$$\begin{aligned}
a_{p,i} = & \, a_{0x1,i}l + a_{0x2,i}l' + a_{0x3,i}F + a_{0x4,i}D + a_{0x5,i}\Omega + \\
& + a_{0x6,i}\lambda_{M☿} + a_{0x7,i}\lambda_{M♀} + a_{0x8,i}\lambda_{M\oplus} + a_{0x9,i}\lambda_{M♂} + a_{0x10,i}\lambda_{M♃} + a_{0x11,i}\lambda_{M♄} + \\
& + a_{0x12,i}\lambda_{M♅} + a_{0x13,i}\lambda_{M♆} + a_{0x14,i}p_a
\end{aligned} \tag{C.7}$$

One has to be careful with the subscripts. The first subscript. here "0" (in general 0 to 4), corresponds to the summation terms in Equations C.4 and C.6 matching those with the same index (0 to 4). The "x" is for the coordinate X, while different coefficients are to be used for Y and s. The third subscript denotes the fundamental argument under consideration (all 14 are listed above), while the last subscript "i" is for the summation, which is different, depending on which sum the argument belongs to in Equations C.4 and C.6.

The fundamental arguments in Equation C.7 are subdivided into the luni-solar nutation terms, given with the Delaunay variables, and the planetary precession terms, with the following equations (Luzum and Petit, 2010):

$$\begin{aligned} l &= \text{Mean anomaly of the Moon} \\ &= 485\,868.249\,036'' + 1\,717\,915\,923.2178''t + 31.8792''t^2 + 0.051\,635''t^3 \\ &\quad - 0.000\,244\,70''t^4 \end{aligned} \tag{C.8}$$

$$\begin{aligned} l' &= \text{Mean anomaly of the Sun} \\ &= 1\,287\,104.793\,048'' + 129\,596\,581.0481''t - 0.5532''t^2 + 0.000\,136''t^3 \\ &\quad - 0.000\,011\,49''t^4 \end{aligned} \tag{C.9}$$

$$\begin{aligned} F &= \text{Mean longitude minus mean longitude of the ascending node of the Moon} \\ &= 335\,779.526\,232'' + 1\,739\,527\,262.8478''t - 12.7512''t^2 - 0.001\,037''t^3 \\ &\quad + 0.000\,004\,17''t^4 \end{aligned} \tag{C.10}$$

$$\begin{aligned} D &= \text{Mean elongation of the Moon from the Sun} \\ &= 1\,072\,260.703\,692'' + 1\,602\,961\,601.2090''t - 6.3706''t^2 + 0.006\,593''t^3 \\ &\quad - 0.000\,031\,69''t^4 \end{aligned} \tag{C.11}$$

$$\begin{aligned} \Omega &= \text{Mean longitude of the ascending node of the Moon} \\ &= 450\,160.398\,036'' - 6\,962\,890.5431''t + 7.4722''t^2 + 0.007\,702''t^3 - 0.000\,059\,39''t^4 \end{aligned} \tag{C.12}$$

$$\lambda_{M☿} = \text{Mean longitude of Mercury / rad} = 4.402\,608\,842 + 2608.790\,314\,157\,4t \tag{C.13}$$

$$\lambda_{M♀} = \text{Mean longitude of Venus / rad} = 3.176\,146\,697 + 1021.328\,554\,621\,1t \tag{C.14}$$

$$\lambda_{M\oplus} = \text{Mean longitude of Earth / rad} = 1.753\,470\,314 + 628.307\,584\,999\,1t \tag{C.15}$$

$$\lambda_{M♂} = \text{Mean longitude of Mars / rad} = 6.203\,480\,913 + 334.061\,242\,670\,0t \tag{C.16}$$

$$\lambda_{M♃} = \text{Mean longitude of Jupiter / rad} = 0.599\,546\,497 + 52.969\,096\,264\,1t \tag{C.17}$$

$$\lambda_{M♄} = \text{Mean longitude of Saturn / rad} = 0.874\,016\,757 + 21.329\,910\,496\,0t \tag{C.18}$$

$$\lambda_{M♅} = \text{Mean longitude of Uranus / rad} = 5.481\,293\,872 + 7.478\,159\,856\,7t \tag{C.19}$$

$$\lambda_{M♆} = \text{Mean longitude of Neptune / rad} = 5.311\,886\,287 + 3.813\,303\,563\,8t \tag{C.20}$$

$$\begin{aligned} p_a &= \text{General accumulated precession in longitude / rad} \\ &= 0.024\,381\,750t + 0.000\,005\,386\,91t^2 \end{aligned} \tag{C.21}$$

C.2. Earth rotation angle

The Earth rotation angle matrix consists of a single rotation around the CIP:

$$\mathbf{R}\,(t) = \mathbf{T}_3(-\theta_{ERA}) \tag{C.22}$$

The Earth Rotation Angle (ERA) accounts for the sidereal rotation of the Earth, being the angle between CIO and TIO and defining UT1 by convention (Luzum and Petit, 2010). Thus, the time variable used to compute the ERA is the Julian day in UT1 with an offset to the epoch 2000 January 1.5:

$$t = \text{Julian Day UT1} - 2\,451\,545.0, \tag{C.23}$$

so that ERA can be computed as:

$$\theta_{ERA}\,(t) = 2\pi\,(0.779\,057\,273\,264\,0 + 1.002\,737\,811\,911\,354\,48t)\,. \tag{C.24}$$

From the EOP the term $\Delta UT1$ is used to obtain UT1 from UTC, which is generally the input time system used.

C.3. Polar motion

The rotational matrix describing the polar motion, which is the difference between the CIP and the IRP is computed via three consecutive rotations:

$$\mathbf{W}\,(t) = \mathbf{T}_3\left(-s'\right)\mathbf{T}_2\left(x_p\right)\mathbf{T}_1\left(y_p\right), \tag{C.25}$$

here, x_p and y_p are the polar coordinates of the CIP in the ITRF and s' is the so-called TIO locator, that "provides the position of the TIO on the equator of the CIP corresponding to the kinematical definition of the 'non-rotating' origin (NRO) in the ITRS when the CIP is moving with respect to the ITRS due to polar motion." (Luzum and Petit, 2010). While x_p

and y_p are quantities subject to measurements and provide alongside with the other EOP, the quantity s' results from a numerical integration of those values. However, it can be approximated using the average values for the *Chandler wobble*, a_c, and the *annual wobble*, a_a, of the pole (Vallado and McClain, 2013), evaluated with respect to TT:

$$s' = 0.0015'' \left(\frac{{a_c}^2}{1.2} + {a_a}^2 \right) t_{TT} \cong -0.000\,047'' t_{TT} \qquad \text{(C.26)}$$